I0805486

JACK MASEY
CONWAY LLOYD MORGAN

COLD WAR CONFRONTATIONS

US EXHIBITIONS AND THEIR ROLE IN THE CULTURAL COLD WAR

LARS MÜLLER PUBLISHERS

This book is for Beverly and Annette

PREFACE

Among the multiple games of shadows that made up the Cold War, the World's Fairs of the 1950s, 1960s and 1970s, together with a range of special exhibitions, provided key opportunities for confrontation between the Free West and the Communist East, each trying to upstage the other through the size, design and contents of their pavilions. From the Kabul Jeshyn Fair of 1956 through to the Osaka Expo of 1970, these events—including notable occasions such as the American National Exhibition in Moscow of 1959—created a formal stage for ideological dispute, echoing the face-off between Nazism and Communism at the Paris Exposition Universelle of 1937.

The story of America's participation in these events is interesting enough in itself as a demonstration of the USA's political and cultural stance at the time, but there is a twist to it. The exhibitions and pavilions were not simply examples of American achievements, but also contained important design and architectural initiatives. The first peacetime government commissions for Buckminster Fuller, Charles and Ray Eames, George Nelson, Ivan Chermayeff and Thomas Geismar, and others, happened in the context of these events.

This book looks at a select number of the international fairs and exhibitions with which the United States Information Agency was involved: the focus is particularly on those where a confrontation between East and West took place on a highly visible stage and with a bewildering array of facets, and in which a significant role was played by architects and designers in shaping America's image during the cultural Cold War. The design and information approaches, the goals and the objectives changed as the dynamics of the Cold War itself changed, or according to the circumstances of the individual event. Each exhibition or pavilion was different in its scale and complexity, each presented a new challenge to the designers involved, each produced a different range of responses, whether from the visitors to fairs or exhibitions, from the US and international press, or from American citizens and lawmakers. These responses were sometimes comic, sometimes moving: there can be no one pattern for experiences that were shared by millions. Nor can there be any exact measure of the influence or effect of events so circumscribed in time and space, but deep and complex in content and context. These endeavors were exciting, the results surprising.

This book began with a chance meeting between the authors in New York some six years ago: it reflects the interest we both shared and share in the Cold War period and the history of twentieth century American design. It is not a memoir, however, nor does it attempt to be a scholarly or exhaustive history of the USIA's role in fair design. It is a compilation of opinions and views—visual and verbal—of a key period in the American hegemony, created from perspectives diffcrent in age and geography, but united by the relevance and excitement of design as a cultural combatant in the confrontations of the Cold War.

Jack Masey
Conway Lloyd Morgan

New York and London, August 2008

A WARM WELCOME TO THE COLD WAR

MARSHALL PLAN TRAVELING CARAVANS WESTERN EUROPE 1948–1951

Du
und
Europa

These German civilians are learning a new word—Marshall—which is part of the description of their new world. After the fall of Nazism, in 1945, there were extensive plans to restore civil government to Germany, partitioned between the four Allied powers: the USA, Great Britain, France, and the USSR. The Russians were looting heavy machinery from German factories under the guise of war damage reparations, and the initial plans on the Western side at first envisaged a demilitarized and even agrarianized Germany. Only in that way would the threat of German warmongering, which had brought war to Europe three times in the preceding century, in 1870, in 1914, and most disastrously of all in 1939, be contained. With an increasing perception of the threat of the Stalinist regime's expansionist policies, this idea was reversed in favor of rebuilding German economic prosperity. The decision by the USA to fund redevelopment in Europe under what was called the Marshall Plan was not simply an economic operation. It was aimed at convincing the people of Germany (and the rest of Europe) that economic progress and democracy went hand in hand. So tents and caravans, trains and barges carried the message of the Marshall Plan wherever they could.

But why was this happening at all? Since the end of World War II, the alliance of necessity between the USSR, and the USA and Great Britain, formed against Germany and Japan in 1942, had been failing. While what came to be called the Cold War did achieve the benefit of no land war in Europe, and no global nuclear annihilation, it was a real, hot war on other fronts, as in Korea, Egypt and the Middle East, and in South East Asia (war claimed 200,000 victims killed between 1945 and 1990). In Europe it remained an economic conflict, a political stand-off, an endless flurry of espionage, and a cultural battleground. The expression "winning hearts and minds" has now become beyond shopworn, but the success of propaganda programs during the War, combined with the lack of a strong democratic tradition in Germany and the presence of powerful Communist parties in countries such as Italy and France, made persuading the people to accept the values of democracy as important as persuading politicians to accept financial support. The cultural territory was as important as the political landscape, if the Russian threat was to be averted.

< Visitors queuing to see the Europa Zug's Marshall Plan exhibits in West Berlin.

This cultural territory was not just the heartland of the intelligentsia, though much effort was expended on both sides to influence the supposedly influential, as Frances Stonor Saunders has shown. There was also an airwave war, addressing a much broader swath of the target populations, and theater and movies (especially in the USSR) were also allied to ideological purposes. These propaganda efforts were not merely directed by the main adversaries against each other, but also at influencing opinion in the developed world and the third world. Here, by the 1950s, Expos, World's Fairs and international trade fairs had become a major focus point for ideological confrontation, a series of opportunities for each side to set out their views of the world, their own achievements and aspirations.

These events were important because of their sheer scale. Hundreds might pack a concert hall to hear Ashkenazy play or a theater to see Nureyev dance. Several thousand might buy and read *Amerika*, the USIA magazine distributed in the Soviet Union, or subscribe to *Encounter*, the British magazine covertly funded by the CIA. But millions visited Expos and World's Fairs, hundreds of thousands saw trade exhibitions and festivals. Visitors were businessmen and government officials, tourists and local people alike. These exhibitions (the USIA was involved in 343 separate showings between 1954 and 1969 alone) were an opportunity to speak to and influence diverse groups, and because of the size of the exhibitions, also a pretext to address a broad number of issues and present a myriad of opportunities for involvement and connection at the same time. Their influence is perhaps, for the very reason of scale, incalculable, but their importance cannot be denied.

MARSHALL PLAN IN ACTION

The illustrations in this chapter show how the Marshall Plan for economic aid to Europe, begun in 1947, was presented to the citizens of Europe through mobile exhibitions, trains, and similar events.

Often working in difficult conditions, in a continent devastated by war, the teams "learnt by doing," and many of those involved were later to be part of the USIA and their efforts to present America and the American view to the rest of the world.

Dutch children watching a puppet show at a Marshall Plan exhibit.

EUROPA

AN OPENING SALVO

In the summer of 1949 an unexpected event happened in an unexpected place. The town of Aspen, Colorado became the host to a festival celebrating the bicentenary of the birth of the German poet and writer Johann Wilhelm von Goethe. The festival was the idea of Walter Paepcke, founder and president of the Container Corporation of America, created in the 1930s in Chicago, which had built a reputation for itself as a company that used contemporary art in its advertising and promotion. The corporation's collection had been exhibited at the Art Institute of Chicago four years before, and Paepcke had helped Walter Gropius, László Moholy-Nagy and other exiles to establish the new Bauhaus in Chicago.

The festival, which lasted nearly three weeks and was attended by 2,000 people, was intended to be a "celebration of the universal man" that would "reunite a shattered Western culture." Paepcke's vision of business as a benevolent moral force was not an isolated one: it was shared by Nelson Rockefeller of Standard Oil, who subsidized business start-ups in Latin America, by Henry Luce of *Time* magazine, and by General Eisenhower, at the time president of Columbia University, who organized the American Assembly. The Assembly brought together businessmen and academics to look at a range of global issues. As Robert Haddow puts it, both institutions, the Assembly and the Aspen Institute, "were aimed at reconciling American foreign policy with a feel-good humanism," by merging "cultural, academic and corporate concerns into a crusade for international peace and prosperity."

One may feel today that these leaders of business were looking at the world through spectacles tinted with a fair degree of rose, but at least they were looking at the world outside their immediate business interests. The Aspen event continued on an annual basis, with a particular emphasis on the role of design. Paepcke had enlisted the support of Herbert Bayer, former Bauhaus teacher and typographer, to develop the cultural potential of Aspen (where he could also indulge in his passion for skiing). Bayer was very involved in the early Aspen conferences and in exhibitions there and in Chicago sponsored by Paepcke. Paepcke also realized—and invested in—Aspen's potential as a ski resort: his interest in the town transformed it. Paepcke's choice of a "good German" as the object of the celebration is an interesting reflection on America's need for Germany as an ally, as well as the European reach of Paepcke's interests.

The Aspen celebration of Goethe was not the only event to mark the bicentenary of his birth. A number of similar events were planned in Germany: the composer Ernst Pepping prepared a new choral work for the Spandau Choir from West Germany to perform in the city of Weimar, where Goethe had lived. But in 1949 Weimar was in East Germany. With both halves of Germany maneuvering for independence—the Federal Republic would be declared that same year, the Democratic Republic a year later—it was no time for shared projects even around such a revered figure. The choral work was premiered in Frankfurt, another city associated with Goethe, amid further celebrations, just as the bicentenary was extensively celebrated in Weimar.

The fact that both capitalists and communists sought to celebrate Goethe illustrates an important facet of the ideological struggles enmeshed in the political, military and cultural fabric of the Cold War, as David Caute has pointed out. Conflict on an ideological basis (as opposed to a religious or political one) is a modern phenomenon, with its roots in the French Revolution and with the conflict between Nazism and Communism as its earlier apex. But there was no shared ground between those two ideologies: even where they might both, for example, loathe modern art, they did so for very different reasons. But both the "free world," and the USSR and the Communist-ruled countries of Eastern Europe, laid claim to the same cultural heritage, the same Enlightenment tradition, and the same concepts of human worth and social progress. Of course the two Germanies shared the same language and history, and so would seek to defend their own definitions of cultural icons such as Goethe, but the shared ground was wider and transcended national boundaries.

This was a situation not without ambiguities on both sides: the way in which the Russians sought to protect the lavish physical evidence of the Tsarist past contrasted oddly with their determination to stamp out the culture and thinking it represented. Similarly in America there were divisions in attitudes towards Europe and "European values"—whatever exactly they were—often voiced by the isolationists and conservatives in both parties. But for all the nuances of understanding and differences over the way forward, the very existence of a common cultural heritage between Russia and the West, which was almost taken for granted, made that heritage a terrain of Cold War engagement.

This common heritage was visible in 1958, when the first International Tchaikovsky Piano Competition was won, in Moscow, by Van Cliburn, a 23-year old from Texas. It has been suggested that the judges approached Khrushchev to ask his permission to award the prize to an American (to which Khrushchev replied, "Is he the best? Then give him the prize!") It has also been suggested that the award was fixed by the Communist Party hierarchy, and given the servility to Party wishes of many cultural organizations, this is plausible. But what is interesting is not what the machinations

Visitors to an exhibit on a barge in Holland.

A tent and Deplirex trucks ready to receive visitors in Italy.

"Marshall … did not talk … about ideologies or armies. He talked … about food and fuel and their relation to industrial production … the relation of production to the organization of Europe and the relation of the organization of Europe to the peace of the world."

James Reston, *New York Times,* May 9, 1947

Le Train de l'Europe receives visitors in the heart of Paris.

were, but why the nationality of the winner mattered at all. What has a music competition to do with the space race, the missile gap, or the future political color of the non-aligned countries?

It mattered because the cultural arena was claimed equally by both sides, and at various levels. There was an almost implicit agreement that winning militarily, winning economically, and winning politically were all important goals, but that winning culturally would also be a major victory. As Wilson Compton, director of the State Department's International Information Administration, said in 1953, "We are relying on armies and armaments to win a 'hot war' if a 'hot war' comes. But winning a hot war that leaves a cold war unwon will not win very much for very long ... winning a cold war may be a practical way of avoiding or largely reducing the needs of continuous gigantic investments in means of 'hot war.'" And a cultural win was not just a highbrow matter, though much effort was expended on this, both with exchange programs between the USA, the Soviet Union, and their various allies, and the hidden funding, on both sides, of conferences and intellectual journals in Europe. It was also important to reach a wider audience on the level of sharing the everyday experiences and aspirations of ordinary citizens. It was in this field that US participation in World's Fairs became critically important, along with special exhibitions such as the American National Exhibition in Moscow in 1959, and other exhibitions in Europe, the Eastern bloc, and the Middle and Far East.

Recent books have drawn attention to specific exhibitions (as with Robert H. Haddow's study of the 1958 Brussels Exposition), or looked at changes in policy and how this was reflected in the Moscow exhibition (in Walter L. Hixson's work), or placed exhibitions in the wider framework of cultural conflict (as with David Caute's book, mentioned earlier). The distinctions that need to be drawn here are between trade fairs, which are normally dedicated to a specific industry or market sector (and as such are outside the scope of this book), and World's Fairs, Expos, cultural exchange exhibitions, and special traveling exhibitions, all meant to promote mutual understanding. One also needs to look at the distinction between "propaganda" and cultural diplomacy, and at the levels of knowledge the different sides had of each other.

Signs for Marshall Plan traveling caravans in Scandinavia.

A Deplirex truck expanded for viewing in Holland.

THE START OF THE COLD WAR

There is a story about a scriptwriter working on a film about the Battle of Agincourt who has the English king declare to his troops: "Men of the Middle Ages, the Hundred Years War is about to begin!" The Cold War was also named retrospectively, and there is much argument about when indeed it began. At the 1945 Yalta or Potsdam Conferences just before and after the end of the war in Europe, as differences began to open up between the wartime allies? With George Kennan's "Long Telegram" from Moscow and Winston Churchill's "Iron Curtain" speech in Fulton, Missouri in 1946, both wake-up calls to perceived Soviet expansionism? Or with the declaration of the Truman Doctrine in 1947, committing America to a world role in defense of democracy? The debate becomes more complex as more material is released from the former Soviet archives. The term "Cold War" had been used in a 1945 essay by the writer George Orwell, though he was talking about the post-Hiroshima world generally, not a specific confrontation. The term was also used, in America, and in reference to the Soviet Union, by Bernard Baruch in 1947: Baruch was a businessman who had also been an economic adviser to Roosevelt. His speech was written by Herbert Swope; the columnist Walter Lippmann picked up on the term and so gave it wide currency.

Early histories of the Cold War, mainly written by Americans, looked at the issue of who had started the war (which in turn affected when it started) but, for later historians, this issue is not the central one. Rather they are concerned with how domestic questions influenced or interacted with foreign affairs, both in the USA and the Soviet bloc (rather than treating home and overseas matters separately), with how changes of leadership could affect policy in both systems, and how perceptions of each side by the other changed and developed. But the 1946 Kennan telegram remains a useful starting point for grasping the question of attitudes on both sides.

Crowds queuing for an exhibit in the Piazza Plebiscito in Naples, Italy.

Deplirex trucks near the Castel Nuovo in the Piazza Municipio in Naples, Italy.

THAT TELEGRAM

In 1946 George Kennan was a 44-year old Foreign Service officer who had been stationed in Moscow from 1933 until 1939. He had specialized in Russian and Eastern European affairs since joining the Foreign Service in 1926. He spoke Russian, among other languages. He returned to Moscow in 1944 as Deputy Chief of Mission. In February of 1946 the US Treasury enquired of the embassy if they could explain why the USSR was reluctant to join the International Monetary Fund and the World Bank. Kennan took this as an opportunity to set out at great length his idea of the Soviet outlook on the rest of the world and the reasons for it. He divided his five thousand word response into five topics: basic features of the postwar Soviet outlook, its background, its projection in practical policy on an official level, its projection on an unofficial level, and, finally, practical deductions for US policy.

Kennan sees the Soviet outlook as grounded in Marxist-Leninist principles, principally the innate opposition between capitalism and Communism, and the inevitability of the collapse of capitalism through internal tensions and war. It is therefore necessary at all times to "advance the relative strength of the USSR as a factor in international society" and to make efforts "toward deepening and exploiting of differences and conflicts between capitalist powers."

Kennan goes on to point out that the premises underlying this official outlook are "for the most part, simply not true." Thus "Soviet party line is not based on any objective analysis of the situation" but "arises from basic inner-Russian necessities." The first of these he cites is the "traditional and instinctive Russian sense of insecurity." This legacy of inadequacy made Marxist dogma, "rendered even more truculent and intolerant by Lenin's interpretation," the ideal regime for Russia, continuously on the defensive against real or imagined enemies without or hidden enemies within. The regime ("a conspiracy within a conspiracy") is unable to accept or respect the idea of objective truth, and so is wholly distrustful of the outside world.

The official promotion of this outlook, Kennan's third section, will be reflected in increasing control in internal matters, and increasing military and industrial activity. In foreign policy they will seek to take any opportunities to enhance Soviet power and prestige. As to international bodies, here again the Soviets will take part if it is to their advantage, or as a means of creating dissent in the West. On the unofficial plane—that is, in areas where involvement would be covert or deniable—Kennan expects the Soviets to be active in promoting their own cause, or in seeking to destabilize the aims of others, and to achieve this ruthlessly, by destruction rather than debate—this in distinction to the punctilious observance of protocol ("if not good manners") in official contexts.

Finally, Kennan considers what strategies and tactics the USA should use in the detailed context he has set out. He begins by pointing out that "we have here a force committed fanatically to the belief that with US there can be no permanent modus vivendi" and that the Russian regime is intent on destroying US society, status and tradition, without recourse to reason or logic. But this is by no means a hopeless situation: comparing the Soviets to the Nazis, he points out that they are "neither schematic nor adventuristic" and they will withdraw if they encounter a strong counter-force: they understand the logic of force. He goes on to point out that the Russian state is not as developed (in industrial or infrastructure terms) as the West, and that the regime is not yet stable, in two major respects: firstly, in terms of the transfer of power (he points out that it took Stalin fifteen years of purges and repression to secure his position after Lenin's death), and secondly, because of popular disillusion ("never have mass of Russian people been emotionally further removed from doctrines of Communist party than they are today.")

It is therefore important, he argues, for the USA to understand carefully and unemotionally the nature of the Soviet regime. It is also important to educate the American public about the true nature of the regime—and here the US Government must take the lead. Thirdly, the health and vigor of American society must be maintained and internal problems confronted. It is also important for the US to convey a positive view of the future, especially to war-weary Europe. Finally "we must have courage and self-confidence to cling to our own methods and conceptions of human society."

Kennan's persuasive arguments rapidly circulated in Washington, where he returned soon after. His proposal came to be called the policy of containment, though that term does not appear in the original telegram. Where it does appear prominently is in an article published in *Foreign Affairs* magazine a year later, entitled "The Sources of Soviet Conduct," and signed by X. It was soon known that Kennan was the author. The Soviet side would have seen the article, but the containment argument was probably already known to Stalin: the telegram had circulated in Washington, and it would have been passed in whole or in précis to Moscow. Stalin's response was to ask his ambassador in Washington to draft his own telegram about the situation in the USA. Rather than writing a factual reply, the ambassador asked his Foreign Minister, Malenkov, to tell him what to say. So Stalin received a set of correct Marxist-Leninist responses unsupported by analysis. Kennan's point about the Russians looking to ideology rather than objective fact was proved.

There are differences between the telegram and the article that go beyond style and audience and the use of the term "containment." There are several omissions: firstly, the remarks in the telegram about the pleasant nature of the Russian people (as opposed to the mind-set of its leaders) are not present. Secondly, the passage at the end questioning the real state of knowledge among Russia's leadership about what was happening beyond its borders is not included. The result is perhaps to undercut the sense of isolation felt by the leadership that is in the telegram. Finally, the need to educate the American people about the real nature of conditions in Russia is also elided. The final proposal, that America can succeed through a policy of containment matched by fair dealings is somewhat weakened.

The policy of containment did not take the form Kennan envisaged. He had advocated a combination of political and economic, as well as military pressures, with clear proof of international goodwill, so as to restrain Russia. However, more hard-line policymakers who wished to encircle and outgun the USSR through superiority of weapons, prevailed. Although Kennan served a number of presidents before turning to an academic career, his influence waned.

The document that embodied American policy in the early Cold War is NSC-68, prepared by the National Security Council in early 1950. This document defines Soviet intentions as "the complete subversion or forcible destruction of the machinery of government and structure of society in the countries of the non-Soviet world." The complexities of Soviet ideology, so carefully analyzed by Kennan, are here brushed aside, and the confrontation takes on a heavily military character. The wording of NSC-68 can be seen as a victory of the hawks over the moderates, and the ensuing policy can be understood as an escalation both of the military contest and of the cultural one, in which, under the Truman Presidency, the military took the lead.

Military development, research, and deployment became also a method of using federal funding to maintain an even pace of economic development: there had been real concerns towards the end of the war that an abrupt transition from a wartime economy to a peacetime one would push America back into the boom and bust cycle of the inter-war years. The Marshall Plan to aid recovery in Europe can also be seen as shoring up European economies to prevent a similar cycle of the economic depression that had led to the rise of totalitarian regimes. The Plan, by creating a market for American goods, would also, it was felt, feed back into the American economy. Those who argue that if Kennan's line of thinking had been followed, the tensions of later years would have been lessened, need to consider also the equally serious consequences of a faltering post-war American economy.

Marshall Plan Caravans could be staffed by as many as 60 people.

Front rows: Caravan Program supervisor Philip George (in bow-tie) with permanent Paris-based staff. Rear rows: local guides hired to staff the Caravan in Naples.

Visitors to a Marshall Plan tent in Holland.

THE FOUNDING OF THE CIA

The Central Intelligence Agency was set up through the National Security Act of 1947, as a consequence of a report presented some three years earlier to President Roosevelt by William Donovan, whom Roosevelt had appointed in 1942 to head the Office of Strategic Services. This office had a dual role: to gather and collate strategic information for the Joint Chiefs of Staff, and to undertake special operations not assigned to other agencies. In his 1944 report, Donovan proposed a postwar centralized intelligence service that would report directly to the President, correlate information gathered by other agencies, and set national intelligence objectives. This new agency would be authorized to engage in covert as well as overt activities. The proposal drew expected howls of protest from the military intelligence organizations, the Federal Bureau of Investigation and the State Department. President Truman, in 1946, set up a Central Intelligence Organization to study the issues surrounding the centralization of intelligence gathering. Their recommendations were broadly in line with Donovan's original proposal, but restricted the new organization to activities outside the USA, and thus formed the basis for the 1947 Act. This was later amended in 1949 to make the CIA's budget invisible and render its information and sources not susceptible to disclosure.

It is the essence of secret or covert operations that they be deniable, and as such, they are the diametric opposite of overt or transparent activities. Both sets of operations may have the same end in view, but to let them become conflated in the minds of the target audience is to devalue both. This is a noteworthy distinction in any examination of government-sponsored activities that have a propaganda component: that is, projects intended to "influence" an audience in some way. This distinction, in its simplest form, leads to a differentiation between "black propaganda," meant to mislead through covert and unverifiable sources, and "white propaganda," meant to persuade through overt, independently verifiable sources. This differentiation was clear to those serving in both the USIA and the CIA who understood that the missions and responsibilities of each Agency were inherently incompatible with the other. It is also a useful distinction when examining USIA's role in Expos or in mounting exhibitions around the world in order that USIA's design policies do not become conflated (as they have been by some) with covert policies or CIA activities.

THE TRUMAN DOCTRINE

After the end of the war in Europe, serious fighting broke out in Greece between Communist sympathizers, who had been active in the resistance to Italian and later German occupation, and those loyal to the Greek monarchy. Great Britain, mindful of her interests in the Eastern Mediterranean, had been supporting the Monarchists, but in 1947 the British government informed President Truman that they could no longer afford to fund these operations.

If Eisenhower and Roosevelt had been skeptical of Churchill's Mediterranean strategy during the War, the importance of denying the Soviets access to the sea was clear to President Truman: a Communist regime in Greece would pose just such a threat. So in March 1947 he asked Congress for funds to support Greece and Turkey. At the same time he announced a much more wide-ranging policy. He promised American support for all free peoples "who are resisting subjugation by armed minorities or by outside pressure." It was an extraordinarily broad commitment. While Truman never referred directly to the Soviets in the speech, preferring to distinguish between democratic and authoritarian regimes, the anti-Communist line he was drawing was clear. This policy, soon called the Truman Doctrine, marked a major shift away from isolationism in American foreign policy, and created an ideological framework for the Cold War alongside the geopolitical aspect covered in the Kennan telegram. Implicit in the Doctrine was not only the offer of force to match force, but also pressure to match pressure.

THE MARSHALL PLAN

Three months after Truman's speech, Secretary of State George Marshall proposed, in a speech at Harvard, that the USA should offer economic support to aid redevelopment in European countries. The offer did not at first distinguish between countries in Western Europe and those under Soviet domination, though the intentions behind the offer were clear. If Europe failed to rebuild her economies, then hunger, poverty and depression would push open the door to Communist influence without Stalin starting any military action. The British and French embraced the policy immediately and organized a meeting of other heads of state to work out ways of implementing the plan. Stalin, who had been caught off guard by the proposal, sent a delegation to the meeting in Paris, then promptly withdrew it, and while allowing Eastern European delegates to stay, forbad them to accept the offer. The result was an American propaganda success. Whether the plan did deliver the economic progress intended, or whether growth would have occurred anyway, is still disputed. But the Plan certainly stimulated the desire for recovery.

When the Plan was signed into law in 1948, Alfred Friendly (who later would become editor of the *Washington Post*) was named head of the new Marshall Plan's Information Program in Europe. Under Friendly's guidance, the Marshall Plan program created a film unit, a magazine unit, and an exhibition unit. Though very small, the Marshall Plan's Information Program would, in time, form the working model for the future United States Information Agency.

When Roscoe Drummond (editor of the *Christian Science Monitor*) replaced Friendly as the program's Information Director in 1949, he hired experienced specialists as chiefs of the separate information units. Peter Harnden, who was an architect and the multilingual son of a US diplomat, worked, while serving as a US officer, on the Nuremberg trials in 1945–46. After the war, he continued to work in Germany for OMGUS (Office of Military Government, United States) by designing exhibits about "US cultural, social and scientific life" which traveled throughout Europe. Some exhibits featured architecture and urban planning as topics. Upon being hired as Chief of the Exhibits Unit, Harnden moved to Paris, established the Exhibits Department, and then hired Philip George to organize and supervise the Marshall Plan's traveling exhibit caravans, tents, trucks and barges. In 1950, the architect Lanfranco Bombelli also joined Harnden's design team in Paris.

The themes of the Marshall Plan exhibits were meant to combat the influence of communism in post-war Europe and were largely focused on promoting the benefits of rebuilding Europe's war-torn nations as modern capitalist societies with good economic ties to the United States. In 1950–51, the "Europe Builds" exhibit traveled to seven European countries, and was seen by a million and a half visitors; in 1952, "The Caravan of Peace" promoted the economic, rather than military, benefits of NATO, and the exhibits in the Europa Zug, "The Train of Europe," addressed the benefits of economic cooperation between Europe and the US.

Philip George, Paris-based American designer, supervised the Marshall Plan's Caravan Program.

"[The Marshall Plan] is directed not against any country or doctrine but against hunger, poverty, desperation, and chaos. Its purpose should be the revival of a working economy in the world."

George C. Marshall, U.S. Secretary of State, June 15, 1947

The Marshall Plan meets the cradle of democracy: The Caravan is set up on the plaza in front of the Olympic Stadium in Athens. The Acropolis is in the background.

> The Europa Zug train takes the Marshall Plan across West Germany.

74765 Nür
EUROP
RIC

"In 1950, a million kids came to the Communist World Youth Festival in East Berlin. We wanted those kids to come to West Berlin to see what western democracy was all about so used the Europa Zug for Marshall Plan exhibits."

Philip George, Director of Traveling Caravans

DOCTRINE AND PLAN

Where the Truman Doctrine and the Marshall Plan overlapped was how the concept of the plan was presented to the people of Europe. In 1948 the Marshall Plan office in Paris was given the responsibility for organizing exhibitions across Europe to support the Plan: Alfred Friendly relocated to Paris to run the organization, where he set up policy guidelines and objectives. Friendly was soon succeeded by Roscoe Drummond, who set about building different media units. He hired a filmmaker, Lothar Wolf, to head the film unit: Wolf was responsible for producing over one hundred documentaries to support the plan.

Drummond hired Peter Harnden to be head of exhibits. Harnden had trained in Army intelligence with Peter Blake (later to work closely with USIA) and at the end of the war in Europe was working for the Information Control Division of the Military Government in Austria. In Vienna, Harnden had met his future wife, Princess Marija Wassiltschikov, a White Russian who had spent the war in Berlin working as a translator in the Foreign Ministry, the agency least infiltrated by Nazism. There she met Adam von Trott and others in the group of anti-Nazis involved with the 1944 plot against Hitler. She was lucky to escape with her life, as her diaries, published many years later, show: they also offer an engaging portrait of life in wartime Berlin.

As Chief of the Exhibits Unit, Harnden was responsible for designing the numerous exhibits that toured Europe, housed in different conveyances: canal barges in Holland, the Europa Zug train in Germany, in "caravan" tents and Deplirex trucks whose walls folded out to make exhibition spaces. These different methods were necessitated by the lack of available exhibition halls in postwar Europe, but also tapped into a European tradition of traveling shows, where fairs and circuses had long brought celebration and color to small towns. Harnden hired Philip George to supervise a number of the traveling caravans. He was already working at the Marshall Plan headquarters as a graphic designer: his life was to move beyond bar charts.

"The show did go on," George recalls. "I was traveling all over Europe. There may have been more than one caravan going on at the same time, and I would watch over them. I would stay with one for three or four days and then go on to another city.

"I was there (with a caravan) for the opening. I laid out the show—in whatever city—and then the staff would take over and erect the tent and open up the Deplirex units, and all that kind of stuff. And, when it was organized—opening day or so—I left and I'd go to the next show. I never stayed with a show." The Delpirex units were French-made trucks whose walls opened out for the display, but some of the units transporting tents were even bigger. "The trucks carrying the tents and exhibit material were large—by European road standards at the time. It wasn't until I was taking the first Marshall Plan caravan to Italy via the Riviera coast, that we came to a series of "S" turns too sharp a radius to make the turn. The French came up with a saving suggestion—jack up the rear wheels of the long trailers, insert a large steel plate on the road bed, pour oil then push the trailer from the side with a small truck—ergo!! We made it by halting traffic both ways that stretched for miles." On another occasion in southern Italy the trucks were too high to go under a railway underpass, so George got the Italians to dig a meter out of the road bed and refill it after the trucks passed (they returned by sea).

Each show would tour for a year before returning to Paris to be refitted. The shows never went back to the same cities twice, so as to generate maximum exposure. The Marshall Plan design teams were very much learning by doing, but building up expertise that the USIA would draw on later: "We were a kind of mini-USIA," George recalls.

THE CREATION OF THE USIA

When Eisenhower was sworn in as President in 1953, he brought with him ideas about government and business cooperation, about the international role of the United States that he had fostered with the American Assembly in the late 1940s, and knowledge of European affairs from his role as Commander-in-Chief of NATO. While accepting the policy of containment and the military build-up enshrined in NSC-68, he also realized that military solutions were not the only option: he had experienced the value of propaganda in World War II. His Secretary of State, John Foster Dulles, did not share this broad approach; feeling that cultural diplomacy was a waste of effort, Dulles believed that a strong military presence was the only argument the Soviets would understand—this despite the fact that the CIA was secretly funding a number of cultural initiatives such as the Council for Free Europe.

Truman and the Congress had five years earlier recognized the value of propaganda. In January 1948, Senator Alexander Smith and Congressman Karl Mundt returned from a tour of Europe to announce that the Soviet Union was "directing a campaign of vilification and misrepresentation" against the Western powers and the USA in particular. It was essential to "counteract the insidious undermining of the free forces resisting communism." Congress supported this proposal and the Smith-Mundt Act became law within a month. This immediately made government money available for funding radio propaganda broadcasts (e.g. Voice of America), for publications, and for exhibitions and cultural exchanges overseas; the Smith-Mundt Act made the dissemination of propaganda within the boundaries of the United States illegal.

In August 1953, Eisenhower established the United States Information Agency; its mission was "to understand, inform and influence foreign publics in promotion of the US national interest, and to broaden the dialogue between Americans and US institutions and their counterparts abroad." The State Department had been involved in cultural activities since the creation of a Division of Cultural Relations in 1938, and after the creation of the USIA retained control of a number of educational programs and of the vestiges of cultural relations. The USIA was an independent agency whose Director reported directly to the President.

Like the CIA, it could only operate outside of the United States: Congress would not have accepted a presidentially-controlled information agency working within the USA. Outside the USA the Agency was also called USIS, the United States Information Service (the acronym was not used within the USA to avoid confusion with the United States Immigration Service).

Presidential support did not mean that the new Agency had an easy ride. Unlike the CIA, the USIA was an overt arm of government and its activities and budgets were always open to scrutiny from Congress and the press. Quite apart from the isolationist element in Congress, there were other conservative legislators who disliked the whole idea of propaganda or who also objected to the sponsorship of programs promoting "mutual understanding" rather than directly promoting US foreign policy objectives. USIA budgets were almost immediately slashed by one third by the Republican Congress that accompanied Eisenhower into power. That so often the Agency had to look to support from the private sector was not merely a reflection of Eisenhower's American Assembly aims but a necessity given the parsimony of Congress.

In 1954, the new President accepted the recommendations of the Randall report with regard to US foreign trade objectives. Internally, the US government was also becoming increasingly concerned about the growth of Soviet influence abroad due to Russian participation in international trade fairs. Eisenhower created the Office of International Trade Fairs (OITF) within the Department of Commerce, and used money from his Special Emergency Fund to set in motion a wider American participation in the growing international trade fair arena.

Thus, with the creation of the OITF and the USIA, the government institutions were in place to implement the Smith-Mundt initiative and carry the Truman Doctrine into the cultural field. A senior Foreign Service officer, George V. Allen (later to be head of the USIA), had in 1948 called the Act "revolutionary" and its consequences "not yet appreciated or even understood by many people." He was about to be proved right.

ATOMS FOR INDIA

"ATOMICS" EXHIBITION, UNITED STATES PAVILION INDIAN INDUSTRIES FAIR NEW DELHI, INDIA 1955

Jack Masey was born in New York. Not long after America entered World War II. Masey was drafted into the Army, soon finding himself working in a camouflage unit whose members included the painter Ellsworth Kelly, the photographer Art Kane, and the fashion designer Bill Blass. The unit spent some time in Leamington Spa, England, training to use inflatable rubber tanks and guns. They took these strange creations into action in France after D-Day.

Returning to America, Masey worked at the New York offices of *Architectural Forum.* Taking advantage of the G.I. Bill that funded education for ex-servicemen, he attended the Yale School of Art and Architecture while continuing to work at *Architectural Forum* during the summer vacations. Shortly after graduating from Yale, he was invited to join the State Department as an Exhibits Officer stationed in New Delhi, India. This seemed an interesting opportunity to travel and see the world. Masey accepted.

It was indeed an opportunity: USIS officers were part of a small team based in overseas embassies and consulates, but, once the USIA was established, the programs were determined by Washington and run in concert with embassy staff. Exhibits officers, because of their design role, were few and so had a freer hand than other officers. India had won its independence from Britain less than a decade earlier, and was an important non-aligned country with a rich cultural and political history. English was an official language, the country was (and is) Asia's largest democracy. Masey soon found himself invited to diplomatic garden parties at Rashtrapati Bhavan, the Presidential Palace in New Delhi. Here he mingled with the Indian Prime Minister, Jawaharlal Nehru, and the Chinese Premier Zhou Enlai, among others. Masey was also interested in Indian art and music: he met performers such as the sarode player Ali Akbar Khan and the sitar player Ravi Shankar.

In his first four years in India, Masey installed small traveling exhibits throughout the Indian sub-continent. His most important project, however, was installing the major exhibition in the US Pavilion at the Indian Industries Fair in New Delhi in 1955. The traveling exhibits allowed him to explore India, while the major exhibition provided the opportunity to meet some of the most important players in the Cold War.

< Jack Masey escorting Prime Minister Nehru on a visit to the Atomics exhibition.

Between Indian independence and 1955, the Cold War had got hot in Korea, with the outbreak of war in 1950, and colder elsewhere, for example with the Russian blockade of Berlin and the Communist takeover of Czechoslovakia in 1948, and the creation of the People's Republic of China after Mao Zedong's victory in 1949. 1949 also saw the formation of NATO, and the first test of a Soviet atomic bomb, four years before the Americans had estimated the Soviets would be able to do so. Despite this heightening of tension, the early 1950s saw increasing prosperity in Europe, and an intensification of propaganda after Congress passed the Smith-Mundt Act at the start of Truman's second term. In 1952 the USA successfully tested a hydrogen bomb, putting it ahead in the arms race, until the first successful Soviet test in November 1955.

In January 1953, Eisenhower began his first term as President of the USA and in March Josef Stalin died, succeeded by Malenkov. However, Nikita Khrushchev was gaining power, and survived an attempt in 1957 to oust him. In March 1958, he emerged as the new leader of the Soviet Union. The confrontation that ensued between Eisenhower and Khrushchev was to be central to the next decade of the Cold War.

Masey with sitar player Ravi Shankar and tabla artist Chatur Lal in the television studio at the exhibition and the celebrated Bharat Natyam dancer, Shanta Rao.

Jack Masey (center) with the USIS Exhibits staff in New Delhi.

Indian construction crews, both
men and women, building
the American exhibition pavillion.

"Whether we want to or not... we can not exist in isolation. Our geography, our history, the present events, all drag us into a wider picture."

Jawaharlal Nehru, Prime Minister of India, September 13, 1955.
On Western and Soviet bloc participation in the Indian Industries Fair.

Ancient building methods and lashed bamboo scaffolding are used to construct the modern concrete, brick and steel US Pavilion.

BUILDING THE PAVILION

The US Pavilion structure was designed by John Vassos, an illustrator and established member of the industrial design community, who had designed a Lucite television set shown at the New York World's Fair of 1939 and, more recently, the first RCA color television camera. The pavilion was built of concrete, brick and steel. It was a single-story building, modernist in appearance, with a giant model of an atom on its roof. At 100,000 square feet (9,300 square meters) it was the largest pavilion at the Fair. Despite the pavilion's modern appearance, it was constructed using traditional Indian building methods.

Children of construction workers look on.

Women as well as men formed part of the construction crews.

> A family group arriving at the American pavilion.

ATOMS FOR PEACE

On December 8, 1953, President Eisenhower addressed the General Assembly of the United Nations, and proposed his Atoms for Peace initiative: peaceful nuclear technology would be passed to other nations through a new United Nations Agency; the United States and the Soviet Union, as the nuclear powers, would each donate a certain quantity of uranium, and provide access to nuclear reactor technology, to the new Agency, which would in turn distribute this new energy resource to suitable member countries.

In some ways this proposal highlights the quiddity of Eisenhower's character as President. Is it to be read as a generous offer to help other nations, a rhetorical flourish (since the Soviets would probably never agree to it), or a cunning ploy to maintain American nuclear supremacy? The American stockpile of uranium was considerably higher than the Soviet one; if each side were to donate the same quantity to the new agency, it would effectively empty the Soviet nuclear larder.

Just as with every other major leader, Eisenhower's decision-making has come under considerable scrutiny. Perhaps what one should bear in mind about Eisenhower's Atoms for Peace proposal, offered in a period of heightened Cold War tensions, is the lack of knowledge each side had of the other: Kennan's long telegram emphasized the ignorance of the outside world about the Stalin leadership, but its obverse was also true. Khrushchev had emerged from Stalin's long shadow as an unknown quantity; a multi-faceted offer such as Atoms for Peace was probably the best way of gaining an informative response. It can also be construed as a win-win position: if the Soviets refused, it would be a victory of opinion for the USA; if they accepted, they would weaken their military potential to the US's advantage.

A number of exhibits on the theme of the peaceful uses of atomic energy had been staged by the USIA in Europe, and it made good diplomatic sense to use a similar theme in India. Because of limited funding, the planning and financing of United States' participation at the Indian Industries Fair was jointly shared by two recently formed US Government agencies: the United States Information Agency (USIA) and the Office of International Trade Fairs in the Department of Commerce (OITF). OITF was responsible for the design and construction of the US Pavilion structure, as well as for the electronics and technology displays inside the Pavilion. For its part, USIA was responsible for installing and operating the Atomics Exhibition, the largest and most politically sensitive component of the US Pavilion, as well as for the programming of a closed-circuit television studio.

"Against the dark background of the atomic bomb ... the coming months will be fraught with fateful decisions ... may they be the decisions which will lead this world out of fear and into peace."

President Eisenhower, Atoms for Peace speech to UN, December 8, 1953

The dome of the American pavilion was surmounted by a red, white and blue model of an atom.

The initial concept plan for the Atomics Exhibition in the US Pavilion in New Delhi was produced by an "Atomic Energy Taskforce" set up in Washington by USIA. Masey was responsible for making the Taskforce's concept plan a reality in the US Pavilion. He would also oversee the construction and operation of the complex Atomics Exhibition.

Among the major features of the Atomics Exhibition were a "magic hands" presentation, in which a trained Indian guide, seated behind a protective wall of glass, demonstrated how radioactive materials could be manipulated by robotic hands; a mannequin, dubbed the "Man from Mars," was dressed in an inflatable "space suit" that visitors were encouraged to touch; a giant thirty-foot-high, full-scale mock-up of the graphite reactor at Brookhaven, New York, gave visitors a sense of the enormous scale and power of the reactor; and a "Hot Laboratory" allowed visitors to listen to, among other things, the gurgling sounds of a Geiger counter.

To welcome visitors to the US Pavilion, the US recruited sixty young Indian guides who were fluent in English, Hindi and at least two other Indian languages. All the guides were graduate students majoring in physics or other scientific disciplines. As a team, the guides—identifiable by their orange Brookhaven jumpsuits—were a key part of the US exhibition, and they made the complex topic of atomic energy comprehensible to a large number of visitors.

Making the use of atomic energy the main theme of the US Pavilion fitted in well with the concept of Atoms for Peace in a number of ways. According to the Chairman of USIA's Atomic Energy Taskforce, Hans N. Tuch, "Eisenhower's Atoms for Peace initiative prompted USIA to make the promotion of peaceful uses of atomic energy a principal objective of its worldwide programming." For visitors to the Fair, it was a reminder that America had created and mastered this technological Janus. Secondly, it was a subtle compliment to the new nation of India to suppose that they also could take part in the new Atomic Age. And the Russians came to see it too.

The graphical information display at the entrance to the exhibition explained the theory and uses of atomic energy.

In November and December 1955, the Communist Party leader Khrushchev and the Soviet Premier Bulganin spent two months visiting India, Burma and Afghanistan. According to Fursenko and Naftali, Khrushchev's intention was as much to learn about the Asian world as to promote Communism. What he did learn was ideologically framed, of course; he described India as being "Kerensky-like," a reference to the Russian government that had briefly held power before the emergence of Lenin and the Communists. Despite finding the Indian Communist Party too sectarian, it would seem he believed in the eventual possibility of a revolution there. The fact-finding aspect of the tour was not his sole objective: he concluded trade agreements with all three countries they visited. And he and Bulganin visited the Atomics Exhibition at the Indian Industries Fair: the honor of showing them around fell to Jack Masey.

A portrait of Albert Einstein
in the information display.

The guides in their orange Brookhaven coveralls in front of a mock-up of a nuclear reactor.

A guide explains the Magic Hands exhibit about handling radioactive materials.

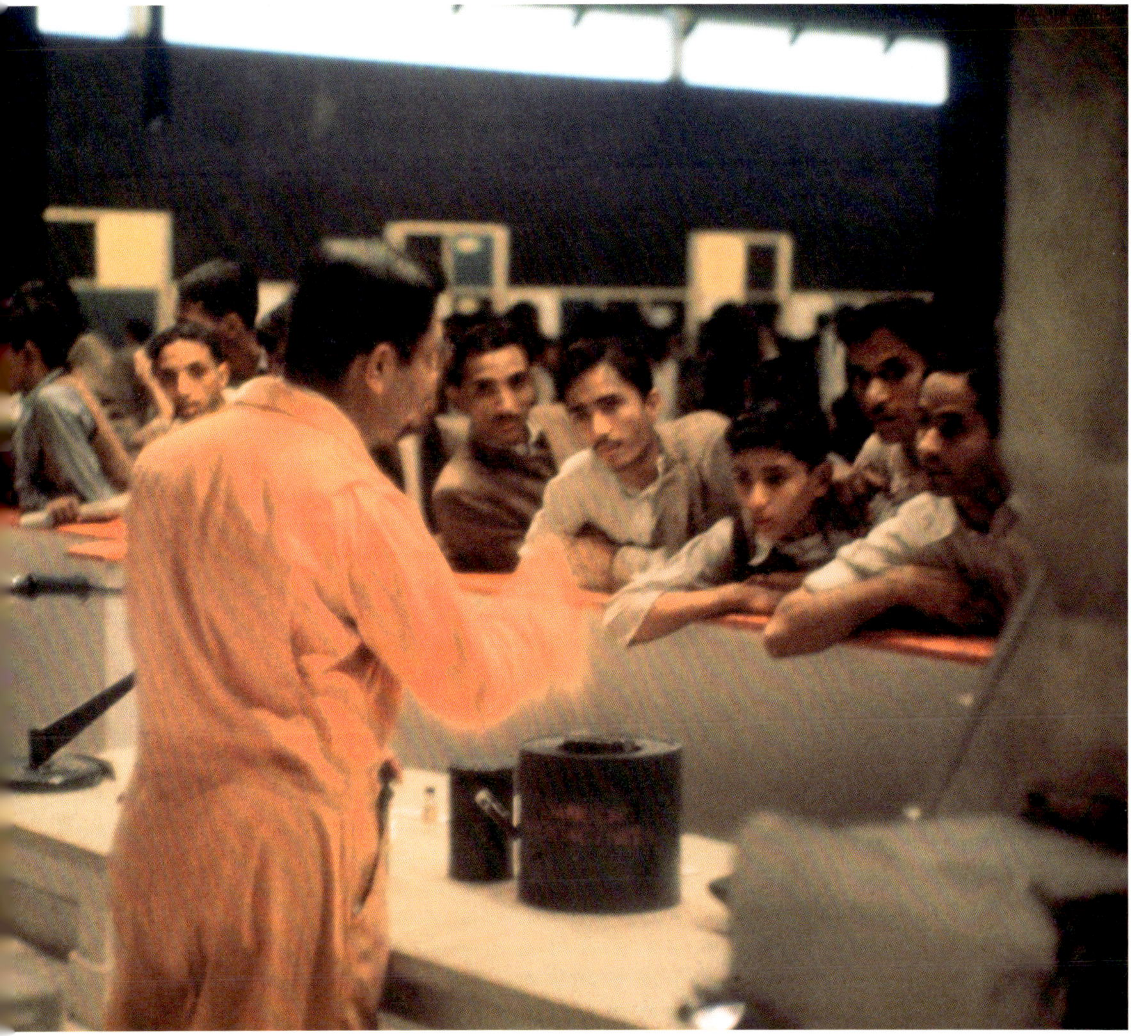

Multi-lingual guides explain a nuclear reactor and the peaceful uses of atomic energy in English, Hindi and other Indian languages.

The exhibition used graphics and bright colors to help inform visitors.

Reactors in the USA

THE SOVIET BLOC AT THE FAIR

The importance of India as a non-aligned nation can be seen in the size of the international pavilions: the major Communist bloc participants (the People's Republic of China, the USSR, East Germany, Czechoslovakia and Romania) erected pavilions occupying a total of 192,000 square feet (17,840 square meters), while the major "Free World" participants (the USA, West Germany and the United Kingdom) erected pavilions occupying a slightly lesser 178,400 square feet (16,570 square meters). The Soviet Pavilion featured a large statue of Lenin, and also displays of mining equipment, tractors, excavating machinery, radial drills, thread-rolling machines, and models of dams. The Soviets also included displays of consumer items that included bicycles, sewing machines, pots and pans, and household goods and appliances.

The Chinese Pavilion used traditional Chinese architecture to house a large statue of Mao Zedong while the Soviet building included a giant statue of Lenin.

The Czech pavilion showed the involvement of Eastern bloc countries.

USSR

GERMAN DEMOCRATIC REPUBLIC

V. LENIN

INDIA AND AFTER

The United States Atomics Exhibition at the Indian Industries Fair was the largest project Masey had so far undertaken. Masey's suggestion to use Indian guides to explain the complex exhibits to visitors in their own language had been a positive benefit.

Once the Fair opened, distinguished visitors were escorted through the Atomics Exhibition including Prime Minister Nehru himself and other Indian dignitaries, and the key visit of the Soviet leaders Khrushchev and Bulganin. These visits emphasized the important role that such

Masey, in his Brookhaven coveralls, shows the Soviet leaders Khrushchev and Bulganin around the American exhibition.

exhibitions were going to play in the wider confrontations of the cultural Cold War; in 1955, the USA participated in thirty-five exhibitions and fairs worldwide as compared to two in 1954. The decision to invest so much in an exhibition in India is further proof of the growing importance of the major non-aligned countries in the political tensions of the time. The USIA, in subsequently declassified documents, can be seen as assessing the attitudes of non-aligned countries, noting the respect many had for the economic developments of Russia and China, but being more concerned with internal development than foreign policy issues. While the USA was not perceived as a military threat, the possibility of atomic war still concerned many.

If the Indian Industries Fair of 1955 was part of a growing pattern of America's serious involvement in international fairs, for many visitors the Fair was an opportunity to relax and enjoy themselves: not all the exhibits were nuclear science, after all. While local Indian press coverage reported that the most popular exhibits at the Fair were the Magic Hands demonstration and the live television broadcasting studio, both located in the US Pavilion, Indian visitors found much to be admired in all the foreign pavilions. At the conclusion of the Fair, a US Embassy report noted that among the most popular exhibits were the "Glass Man" in the East German Pavilion, a BOAC model airplane in the UK Pavilion, a profusion of realistic wax fruits in the Chinese Pavilion, bottles of vodka and tins of caviar in the Soviet Pavilion, and hand-made musical instruments in the Czechoslovakian Pavilion.

1956 was to see a major hardening of the Cold War, not over the usual locus of Berlin, but in the abortive attempt by the British and French to prevent Egypt's nationalization of the Suez Canal, and the brutal repression by Soviet military forces of Hungary's attempt to free itself from Communism. For the USIA and Masey, the year was to see another challenge, in Afghanistan.

A SPLENDID PLEASURE DOME

UNITED STATES PAVILION
JESHYN INTERNATIONAL FAIR
KABUL, AFGHANISTAN
1956

Kabul is not a premier location on the map of mid-20th century design. Yet, in the year 1956, the capital of Afghanistan had its moment. The immediate pretext was the Jeshyn International Fair, and the wider context was the Cold War.

In 1956, the OITF had begun to commission architects and designers from the private sector to design US exhibitions at overseas fairs. Under this arrangement, Raymond Loewy's chief collaborator William Snaith designed a new building and exhibit in Damascus, Finnish-American architect Reino Aarnio designed the Stockholm Fair building, and Raymond Spillman did likewise at the Zagreb Fair.

The Kabul Fair was a different proposition. The US Government decided to take part very much at the last minute: the US Embassy in Kabul had been lobbying for American participation, given the presence of Russian, Chinese and Czech pavilions, but the State Department only agreed to participate a few months before the opening date, perhaps prompted by a now declassified intelligence report that detailed the sizes of the Communist pavilions. This clearly brought the Jeshyn Fair under the terms of a classified agreement, made a year earlier, justifying greater US participation in fairs "when major Soviet bloc competition is present." The USIA asked Masey to come up with a proposal.

Masey quickly realized that the only person who would be likely to meet the time deadline was Buckminster Fuller, whose geodesic structures had often been featured in *Architectural Forum*. Masey flew back to the US, met Fuller in New York—"Call me Bucky" was Fuller's opening remark—and asked him to design and build a hemispherical dome, 100 feet (33.5 meters) in diameter, within six weeks. Bucky said yes, he could.

Few in Washington would have heard of Fuller or taken the risk of hiring him, as his career to date had not been entirely successful, and he had never built a dome as big as the one proposed. At Masey's urging,

< Afghan workers assembling the Buckminster Fuller dome in Kabul.

Signs designating construction areas at the Fair: later used as a cover for ID Magazine.

the US Government gave Fuller the order to proceed on May 25, 1956. Fuller's dome was designed, calculated and engineered in seven days, and manufactured in thirty days. After two weeks of testing in Raleigh, the dome was loaded onto a DC-4 aircraft and departed for Kabul. The dome components were flown to Kandahar, Afghanistan, where they were loaded onto two smaller aircraft for the last lap of the journey to Kabul. (The DC-4 could have landed in Kabul, but there was not enough airplane fuel in Kabul to allow it to return to Kandahar.)

In forty-eight hours—from July 19 to July 21, Jefferson Davis Brooks III, Fuller's construction engineer, supervised the erection of the geodesic dome on the US site in Kabul. Brooks was assisted by local Afghan workers. According to Fuller, the King of Afghanistan was so impressed that he asked for the dome as a gift. The State Department refused. The dome had already been scheduled for use elsewhere.

On a local level, the US did perhaps win the propaganda war. The Soviets brought in 200 technicians to build their pavilion; the Chinese erected a huge statue of Mao, while the Afghans erected the US Pavilion themselves. They could relate to the structure's architecture through their own vernacular yurt tradition, even if the materials were different. In addition, Masey, together with Tom Hall Miller, the US Pavilion manager, employed Afghan guides to explain the exhibits within the dome, such as the Talking Chicken and Talking Cow, the Bouncing Ball Bearing and the Lionel model train set, in the local dialects of Afghanistan. Outside, there were displays of agricultural implements such as tractors and ploughs, and, at night, Cinemascope film showings.

The new OITF policy was discussed in the February 1957 issue of *Industrial Design* magazine: the main theme was a feature on US Pavilion design at international exhibitions. Kabul was included along with Stockholm, Salonika, Izmir, Damascus, Zagreb, and Vienna. Each US Pavilion got a short review featuring portraits of the designers and their comments on

the design challenges of each Fair. The author of the main article was Jane Fiske Mitarachi, who drew particular attention to OITF's decision to hire design professionals to design US pavilions around the world. Mitarachi called the International Trade Fair Program "undoubtedly one of the most serious responsibilities that the design profession has ever assumed." She felt that the government as client "did look to designers for help in this area and got some measurable results."

Mitarachi is clearly seeking to promote a wider role for designers, and in so doing she glosses over the fact that Masey was a US government employee (as was Bernard Pfriem, designer of the US Pavilion at Salonika) and that the pavilion at the Kabul Fair was never part of the tendering process. "In throwing the problem open to designers," she wrote, "it [the government] not only gave members of the profession a good chunk of government work, it also gave them a new kind of responsibility for presenting America, and in some cases, for formulating America's approach to other nations ... Designers were being asked to be propagandists: design became a tool of communication and the visual composition of an exhibition a secondary problem."

She also criticizes the OITF as a client for not giving designers enough information about the host countries and not enough time for research. She suggests the designers be more involved in the choice of material on display, and that in the future more clarity about the criteria for selecting designers and evaluating results would be desirable. Her criticisms were well taken. Within the US government, memoranda mirroring Mitarachi's comments had been circulating between USIA and the Commerce Department since the end of the Kabul Fair. By the time Mitarachi's article appeared in print, USIA had already established its own professional exhibition design office in Washington.

Buckminster Fuller and Jack Masey with a model of the Kabul dome.

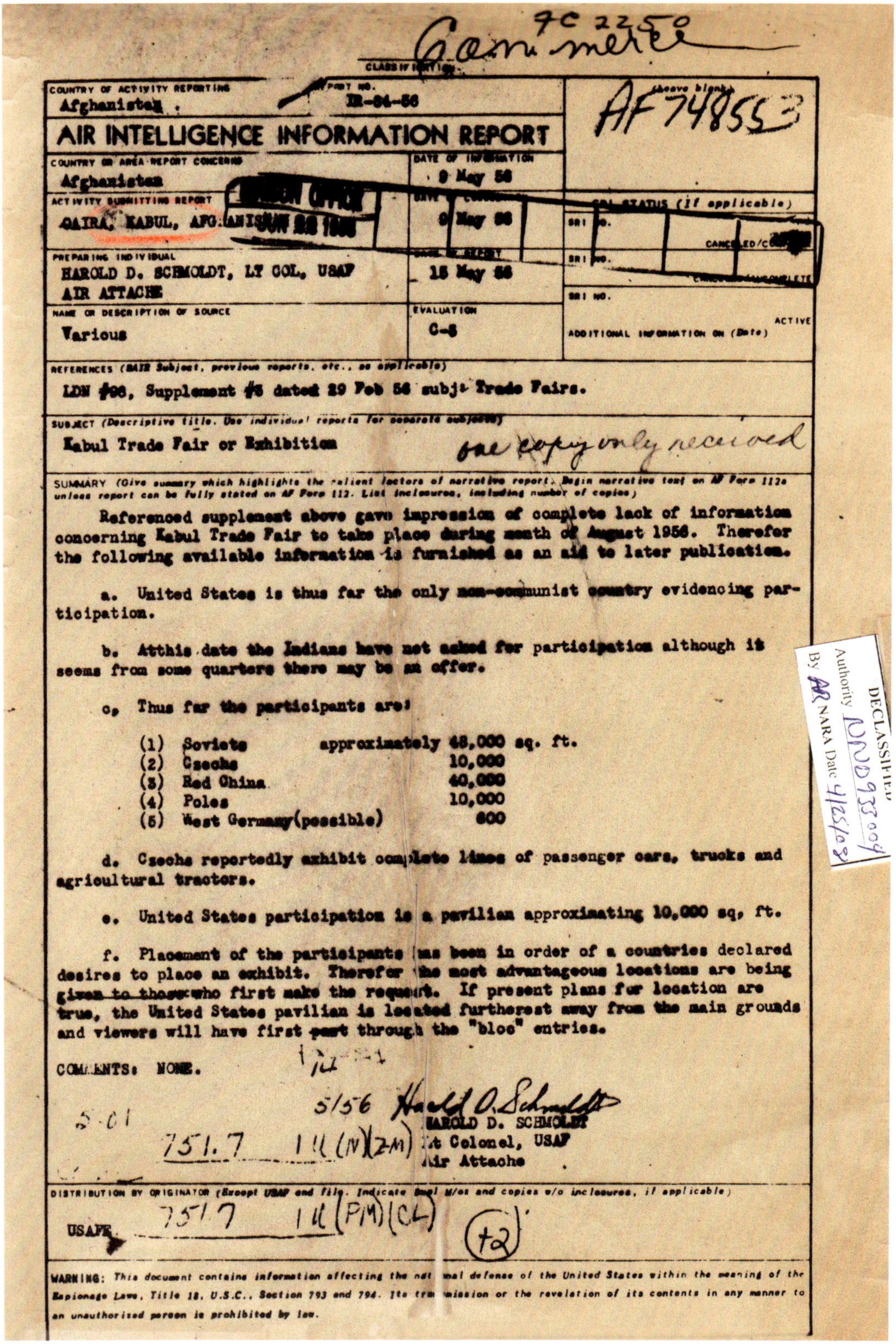

Commerce
7C 2250

CLASSIFICATION

COUNTRY OF ACTIVITY REPORTING	REPORT NO.	(leave blank)
Afghanistan	IR-84-56	AF 748553

AIR INTELLIGENCE INFORMATION REPORT

COUNTRY OR AREA REPORT CONCERNS: Afghanistan	DATE OF INFORMATION: 9 May 56	
ACTIVITY SUBMITTING REPORT: OAIRA, KABUL, AFGHANISTAN	DATE OF COLLECTION: 9 May 56	SRI STATUS (if applicable) / SRI NO. / CANCELED/COMPLETED
PREPARING INDIVIDUAL: HAROLD D. SCHMOLDT, LT COL, USAF AIR ATTACHE	DATE OF REPORT: 15 May 56	SRI NO. / CANCELED/COMPLETE
NAME OR DESCRIPTION OF SOURCE: Various	EVALUATION: C-6	SRI NO. / ACTIVE / ADDITIONAL INFORMATION ON (Date)

REFERENCES (BAIR Subject, previous reports, etc., as applicable)
LDN #96, Supplement #5 dated 29 Feb 56 subj: Trade Fairs.

SUBJECT (Descriptive title. Use individual reports for separate subjects)
Kabul Trade Fair or Exhibition — one copy only received

SUMMARY (Give summary which highlights the salient factors of narrative report. Begin narrative text on AF Form 112a unless report can be fully stated on AF Form 112. List inclosures, including number of copies)

Referenced supplement above gave impression of complete lack of information concerning Kabul Trade Fair to take place during month of August 1956. Therefor the following available information is furnished as an aid to later publication.

a. United States is thus far the only non-communist country evidencing participation.

b. Atthis date the Indians have not asked for participation although it seems from some quarters there may be an offer.

c. Thus far the participants are:

(1) Soviets approximately 48,000 sq. ft.
(2) Czechs 10,000
(3) Red China 40,000
(4) Poles 10,000
(5) West Germany (possible) 600

d. Czechs reportedly exhibit complete lines of passenger cars, trucks and agricultural tractors.

e. United States participation is a pavilion approximating 10,000 sq. ft.

f. Placement of the participants has been in order of a countries declared desires to place an exhibit. Therefor the most advantageous locations are being ~~given to those~~ who first make the request. If present plans for location are true, the United States pavilion is located furtherest away from the main grounds and viewers will have first ~~part~~ through the "bloc" entries.

COMMENTS: NONE.

5/56 Harold D. Schmoldt
HAROLD D. SCHMOLDT
Lt Colonel, USAF
Air Attache

751.7 1 ll (N)(2M)

DISTRIBUTION BY ORIGINATOR (Except USAF and file. Indicate Encl M/os and copies w/o inclosures, if applicable)
USAF 751.7 1 ll (PM)(CL) (+2)

WARNING: This document contains information affecting the national defense of the United States within the meaning of the Espionage Laws, Title 18, U.S.C., Section 793 and 794. Its transmission or the revelation of its contents in any manner to an unauthorized person is prohibited by law.

DECLASSIFIED
Authority NND933009
By AR NARA Date 4/25/08

Air Intelligence Information Report issued by the American Embassy in Kabul, noting the size of Soviet bloc presence at the Jeshyn Fair. This encouraged US involvement in the Fair.

WORKING FAST

Jack Masey arrived in New York to make arrangements for the Kabul pavilion on Sunday May 20th, 1956: the Fair was set to open on August 25th. He kept a log of the flurry of meetings that took place over the next hectic six days.

On Sunday he met Harrison McClung, Director of OITF, who proposed a tent: "Tent ugly, particularly by contrast with anticipated slickness of Iron Curtain pavilions," was Masey's comment. They went on to discuss budget and staffing

On Monday, Masey contacted the USIA officer in Berlin to get a quote for a prefabricated structure, and contacted an American maker, Unistrut. He called *Architectural Forum* for Buckminster Fuller's number, and phoned him. He also spoke to Robert Warner with OITF in Washington: "... told him that I could hardly be expected to design exhibit without knowing area of US site or amount of money." He also met with Howard Messmore in OITF Procurement to obtain a list of potential exhibits with photos and area dimensions. That evening McClung phoned him at home to discuss the Fuller idea and the German pre-fab: "He was particularly excited with the geodesic dome."

On Tuesday, a further round of meetings revealed that the budget for the pavilion was $30,000 "and in the case of the geodesic dome could be upped." On Wednesday, Bucky called to postpone their planned meeting till the next morning. Masey called Warner: "... pointed out the exigency of the situation ... we must use Bucky's dome." Later, OITF Procurement came up with a list of exhibits: "... the same old crap—sewing machines, talking chickens, bouncing ballbearings etc."

On Thursday, Masey and Bucky met for breakfast. "He pulled out section drawing of dome, approximately 110 foot diameter." And an estimate of cost, excluding transport of $35,270. "I was excited as hell." Masey flew to Washington that morning, met that afternoon with Warner, who suggested a $5,000 circus tent. "It was grotesque and he knew it." They then went to meet McClung, who "was excited about the dome and after some tight-rope walking, he bought the idea." Masey called Bucky with the news: Fuller sent one of his engineers, James Fitzgibbon, to Washington.

On Friday, in Washington, "Fitz showed large photo-blowup of dome, as well as sample of skin material. I was overwhelmed by both." They spent the rest of the day discussing details before the contract was handed over to Fitzgibbon. The notes end: "... as far as I know, we are going to have a dome!"

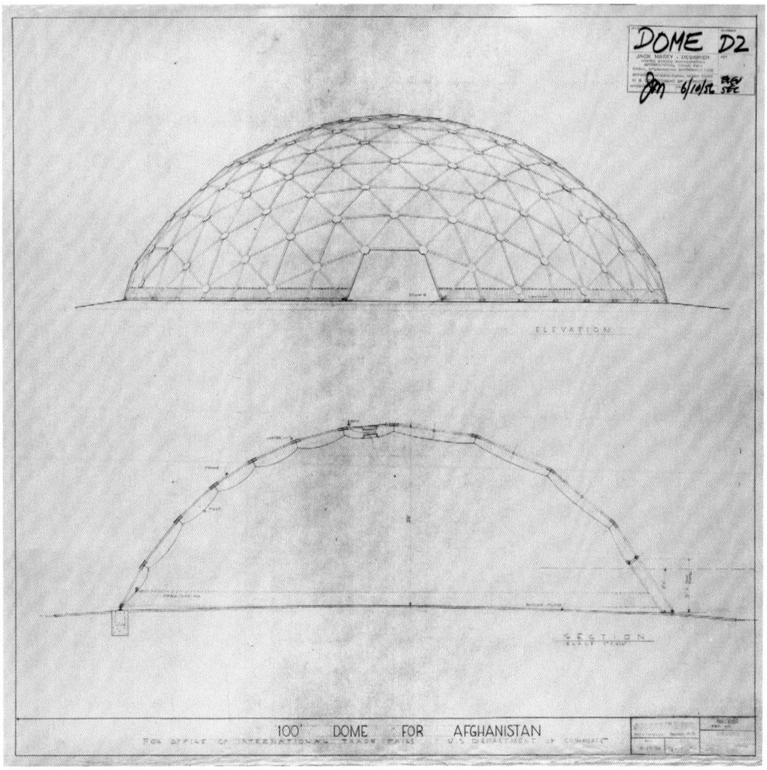

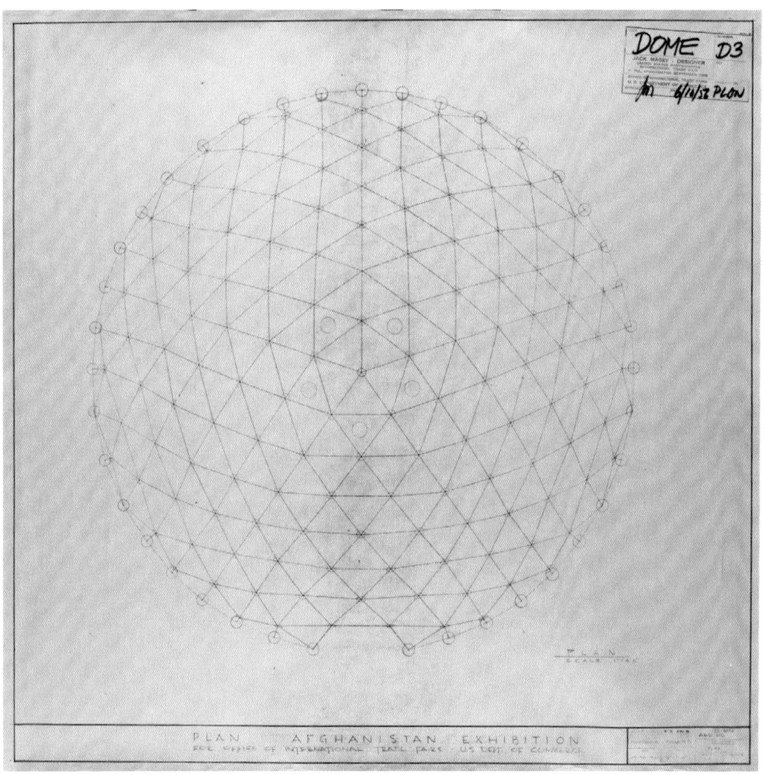

Section and exterior elevation and overview of the roof structure of the Kabul dome.

THE DOME EFFECT

Another reflection on the dome in Kabul came in an article in the *Nation* by David Cort, published in March 1958. Cort was a *Time* journalist who had been hired in 1936 to be Foreign News Editor with *Life,* and who also wrote occasionally for other magazines. His article singled out the individual efforts of Buckminster Fuller and Jack Masey, and contrasted their innovative solution to both the overblown efforts of the Soviets and Chinese, and the lumbering committee-led efforts of the Department of Commerce, especially at the 1957 Zagreb Fair.

"A miracle was required," wrote Cort. "The growing American philosophy is that miracles are produced by corporations. But what was wanted was far beyond any existing American corporation to deliver." He points to the success of the dome, not the exhibits: "Every testimony was that this last-minute victory in the charm war was won by the dome, not by the merchandise inside." In conclusion, he lambasted the food display at the US Pavilion at the Zagreb Fair ("It had no mark of brain, spirit, abstraction or hope: it simply presented a marvellously refined bestiality of the belly. Yet this is the level of our foreign policy, because it is also the level at which corporate advertising works.") He draws attention to the role of individuals: "... we were once a masculine, ascetic, roving adventurous people who created new solutions as fast as new problems arose." One of the first articles that Cort had published after joining *Time* was a piece on Fuller and they kept in touch thereafter. It is not fanciful to suggest that his account of the Kabul Fair came directly from Buckminster Fuller and his team—Cort does not allege that he attended the fair himself, nor that Bucky did, but gives certain details that only an eye-witness would know.

With this somewhat romantic view of American history, Cort can be seen as aligning himself with those, like Henry Luce, who believed in free enterprise untrammeled by government. This view stands in contrast to the opinions of, for example, Walter Paepcke and his Aspen circle, who believed in a more humanist and inclusive approach to the developed world and the Third World. Both groups accepted the need to confront Communism abroad as well as America's duty to play a leading role in the world. They disagreed about whether aid should accompany trade, or whether unfettered capitalism alone would bring democracy to the globe: this in turn was part of a wider disagreement about the nature of foreign policy, not limited to trade fairs and Expos.

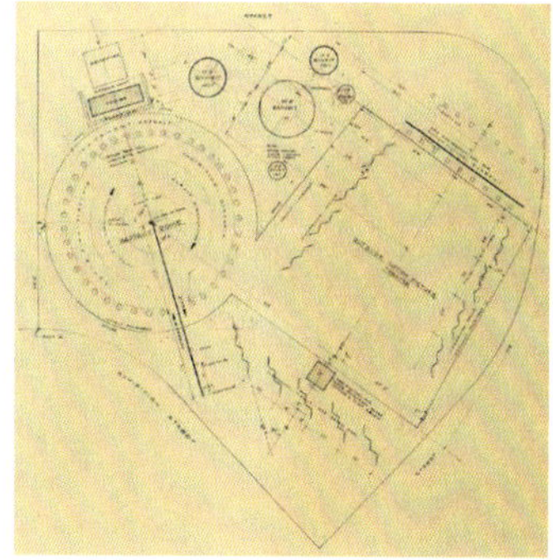

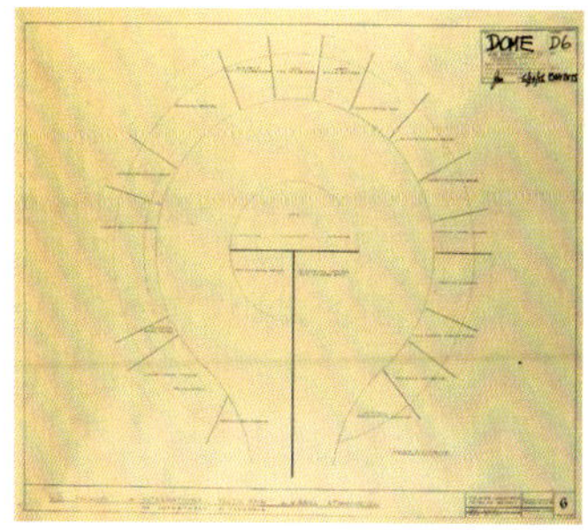

Plan of the US site.

Plan of the interior layout.

> The Fuller dome, in sections, delivered by two aircraft to Kabul Airport.

YA-AAC

CONSTRUCTION

While some press reports talk of “American engineers” in Kabul, the only American engineer there was Jefferson Davis Brooks III, Fuller’s engineer who had accompanied the dome to Kabul to supervise its erection. He was assisted by John Dixon, Fuller’s personal assistant. Additional construction help was initially provided at the US site in Kabul by a team of six Austrian workers who had previously put up the Austrian pavilion.

The local infrastructure was primitive—there were less than 9 miles (15 kilometers) of railway in the country. The Russians and Chinese had brought in large teams of their own engineers, 200 of them in the Soviet case. Part of the success of the dome was perhaps due to the fact that the Afghans helped to erect it themselves.

Jefferson Davis Brooks III (left), Ambassador Sheldon Mills (right) and Afghan official insert the final strut to top out the dome.

Masey (left) and John Dixon oversee site preparation.

Workers assembling color-coded hubs.

BUCKY'S VIEW

Buckminster Fuller's own account is from his book *Critical Path.* Bucky did not only use his engineering skills in Afghanistan, he also deployed his immense arsenal of knowledge. He realized that the country was "the heart of the Heartland," using the term in the sense defined by Mackinder at the start of the twentieth century.

Sir Halford John Mackinder is, alongside Alfred Mayer Mahan, one of the inventors of the concept of geo-politics, analyzing sea power and military force, not just in terms of trade and politics, but as bound up with physical geography. His theory was that whichever nation extended its political control over the Eastern European gateway to the Eurasian landmass would ultimately gain mastery of the entire globe. "Who rules East Europe commands the Heartland; who rules the Heartland commands the world-island; who rules the world-island controls the world." For Mackinder, as for later strategists, the heart of the heartland is Afghanistan, the barrier that prevents Russia from directly accessing the Indian Ocean. Fuller draws attention to Mackinder's theory in *Critical Path* when writing about the chimera of world dominance. He also describes his involvement in the Kabul Fair in somewhat dramatic terms. He suggests, for example, that it was the British Embassy in Kabul that, via the Foreign Office, alerted the State Department to the intentions of the Russians, Chinese and others to participate in the Jeshyn Fair. From declassified sources, it is now known that it was the US Embassy in Kabul that pressed for involvement by the USA.

The nylon fabric skin being attached to the dome structure after erection is complete.

> Finally complete, the dome is ready for installation of the interior and exterior exhibits.

>> Visitors arriving at the dome for the exhibition.

INSIDE THE DOME

Despite Masey's comments on the contents, the US Pavilion in Kabul did in fact contain a number of agricultural exhibits that would have been of interest to the local people. There was also a working television studio that generated a lot of interest, as did the giant outdoor Cinemascope film shows every night.

Translucent dome skin bathes interior exhibits in a golden glow.

The dome with Hindu Kush mountains in background.

Entry panel with New York skyline and message of welcome from President Eisenhower.

The mechanical cow and other items inside the dome.

Visitors to the dome.

Afghan guides explain television studio to visitors and a lathe.

> The Lionel model train set and visitors.

SOUNDINGS

Was the US Pavilion in Kabul, then, a success? The USIA sent in an opinion poll expert, Emily Jones, to organize a survey of visitors' opinions. As Ms Jones has said, "It was probably the first and last time anyone consulted the Afghan people on their views." The survey was undertaken by law students from the University of Kabul, and consisted of a "before and after" survey, to ascertain whether the US effort had made a difference to Afghan opinions of America and, in addition, which of the pavilions had been most impressive. As Ms Jones said, "It was recognized that it was sometimes easier to get an idea of a person's opinion on a subject by asking about it indirectly, through the images of the fair, than by a direct question." But the results of the survey are inconclusive, partly because Afghan Government officials tried to intervene, either by obstructing those conducting it, or by spreading rumors that the questions were about religious beliefs or about Afghan views of their own government.

If there was one key statement that the dome made, it was that the structure was completely modern and completely American. This architectural statement at Kabul was as important—if not more important—than the contents, which consisted to a great extent of whatever was available at the time.

For an exhibition that can almost be called unplanned, given the three months between commissioning and opening, it was a more than considerable success. It also highlighted the potential for a strong architectural statement as a way of distinguishing the American entry as original and exciting. In some ways, the Kabul dome was a quintessentially Cold War object, created for a specific confrontational purpose. The dome was a symbol of American technical expertise, but also of new ways of thinking about the world of international exhibitions.

Drawings used in USIA opinion poll to measure visitors' reactions to Western and Soviet bloc Pavilions.

> The Cinemascope film and the translucent dome lit up the Kabul night.

“The building… went up in two days and it is a beautiful building to handle. When you light it inside at night you just get this big glow, and it reminds me of that corny expression ‘out of this universe.’”

Tom Hall Miller, Manager, US Pavilion, Jeshyn International Fair

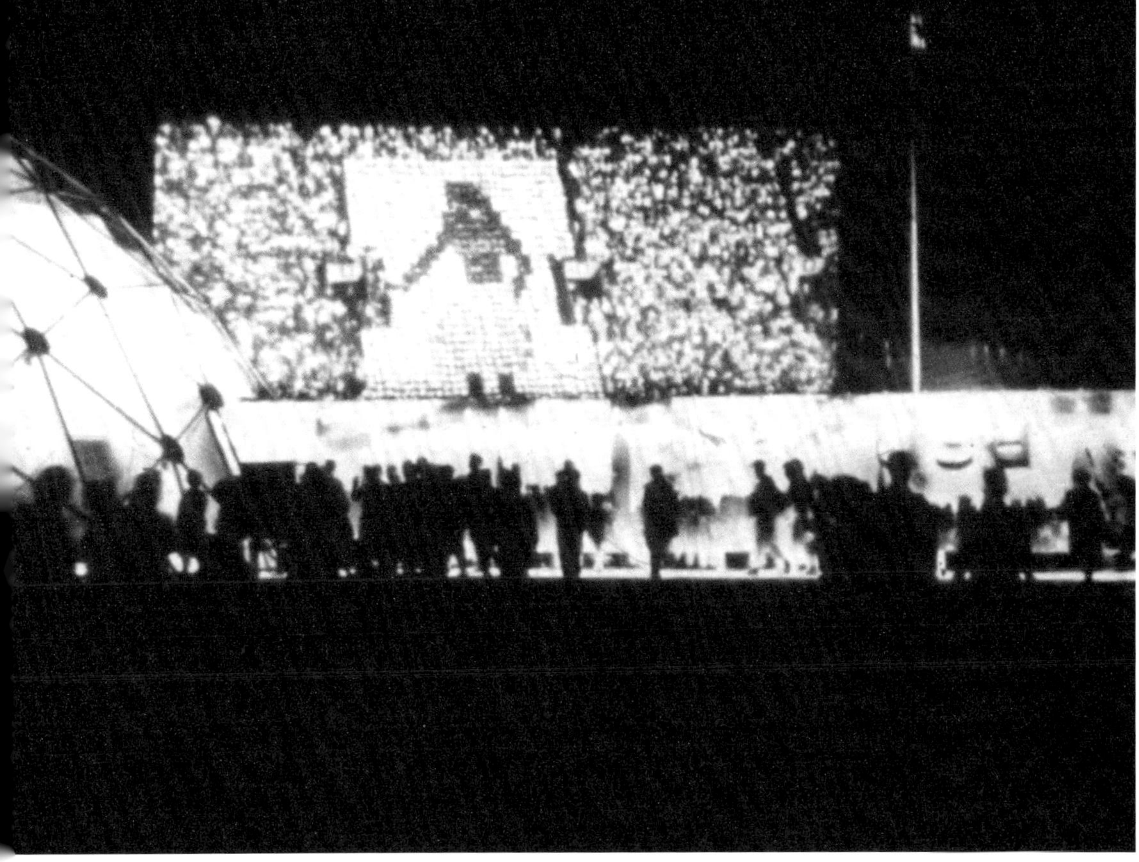

WALLS
IN BERLIN

UNITED STATES EXHIBITIONS
GEORGE C. MARSHALL HOUSE
WEST BERLIN
1957–1959

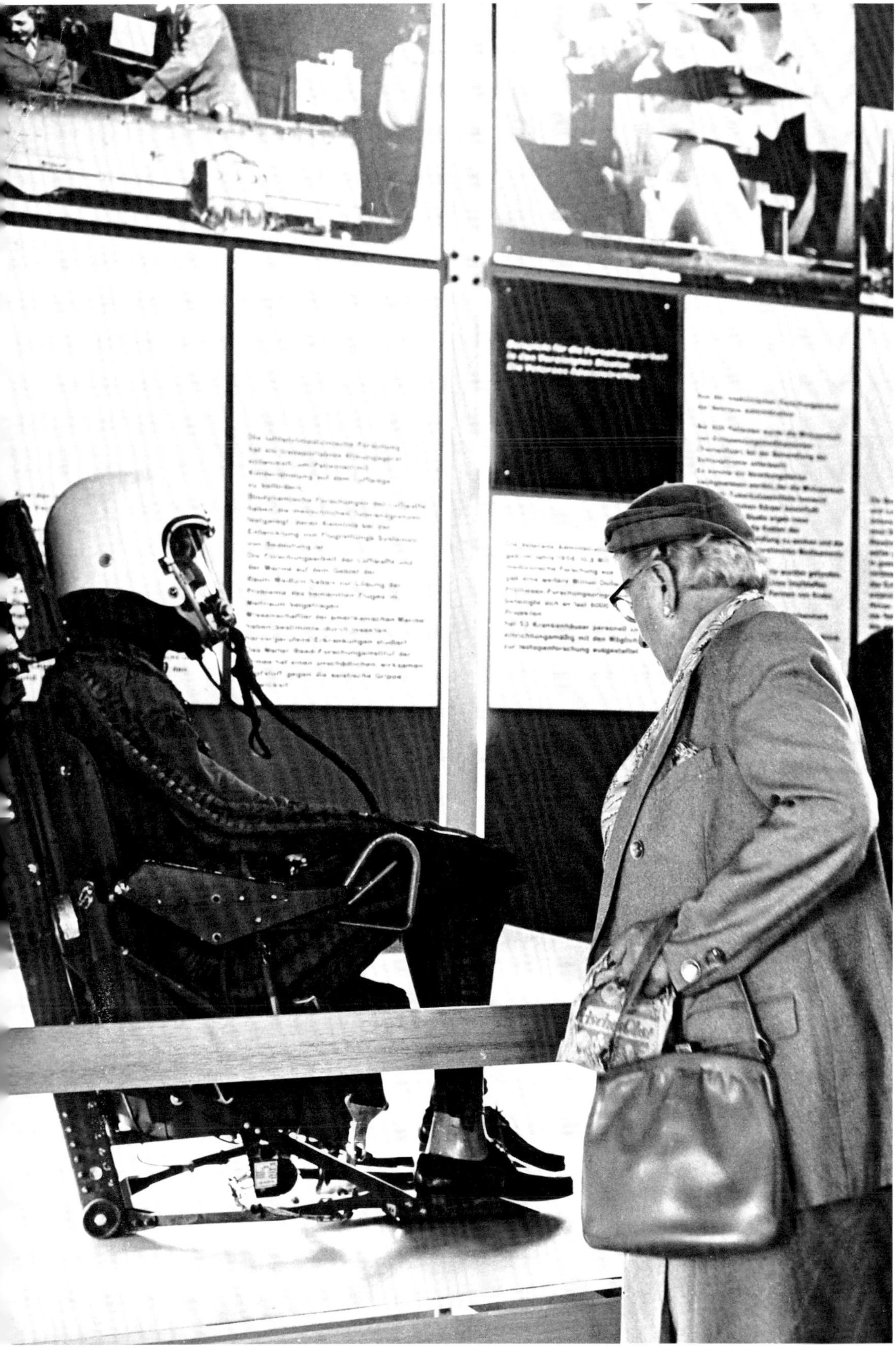

If the cultural Cold War had a front line, Berlin was on it. The blockade of the city had been one of the first clashes of the Cold War, Khrushchev's Berlin ultimatum of 1957 had again raised the temperature there, and the later Berlin Wall was for long to be the symbol of the Cold War itself.

The division of Berlin into four zones, American, Russian Soviet, British, and French, mirrored the partition of Germany at the end of the war between the victorious allies. But after the establishment in 1947 of the Federal Republic (West Germany) and the Democratic Republic (East Germany), Berlin remained a divided anomaly. It was a source of affront to the Soviets, this island of capitalism within their sphere of influence. For the others, it was both the tangible reminder of their victory and an ideal observation post for looking over the enemy's shoulder, a base for espionage and intrigue, a testing ground of Cold War tactics. It is not surprising that over a third of Marshall aid money for Germany was spent in Berlin.

In October 1950, the George C. Marshall House, a large exhibition hall designed by Bruno Grimmek, a municipal architect, opened in West Berlin's Exposition Park. Though architect Peter Blake described it as "looking like a huge, abandoned automobile showroom," the Marshall House was to be the site for a number of USIA exhibitions.

At the end of 1956, upon the completion of tour in New Delhi, Masey was assigned to USIA headquarters in Washington. There, he served with the Agency's Exhibits Division as planner and design director for a number of American exhibitions that—over a two-year period—would reach audiences in Western Europe, Latin America and Asia.

From late 1956 to mid-1958, Masey worked with a diverse and up-coming group of American architects, industrial designers, museum curators, and editors to create special thematic exhibitions depicting various aspects of American life and culture for circulation abroad. Among his priority assignments in this period were exhibitions scheduled for installation in the Marshall House in West Berlin.

During the Berlin Airlift, the city had become, in the minds of many, an oasis of freedom located deep inside Soviet-occupied Germany whose welfare depended on the non-communist world. As such, Berlin, in the years before the construction of the Berlin Wall, was the point at which East met West.

< German visitor examines a space medicine exhibit at Marshall House.

It was the city that, more than any other in the world, offered each side the best opportunity to make cultural, economic or personal contact with the other.

USIA's decision to stage an exhibition on American architecture in Berlin during the 1957 Interbau trade fair and architectural event was from the first, therefore, a politically charged one. It was also a high-profile moment, as Interbau had invited a number of well-known architects, including Le Corbusier, to build projects for the occasion. Masey was asked to direct the design of the US exhibition. For the contents, he contacted, not without some misgiving, Peter Blake, whom he had met while both had worked on the staff of the *Architectural Forum.*

Peter Blake was born in Germany, the son of a prosperous lawyer who had served in the Kaiser's army in World War I. In the 1930s, his family fled Germany to Britain, where Blake went to school, and began studying architecture while working for Serge Chermayeff (also an émigré but from tsarist Russia). When war broke out in 1939, Blake volunteered for the British army, but was refused as an enemy alien. He then continued his studies in the US, working at the same time for the *Time*-owned magazine *Architectural Forum.* He served in the US Army as an intelligence officer. After the war, Blake worked at *Architectural Forum,* and then was appointed by Philip Johnson to be nominal head of the Department of Architecture and Industrial Design at the Museum of Modern Art in New York. He worked with Philip Johnson on a number of projects until 1950 when he rejoined *Architectural Forum.* There, he first met Jack Masey, and the two took an instant dislike to each other.

Blake, too, took Masey's call with some misgiving, but he was being offered an opportunity to revisit his home city, and the prospect of working on an exhibition. He accepted, and flew out to Berlin to visit the Marshall House site. On arrival, he recounts, "there was Masey, fairly scruffy as always and otherwise unchanged also." He then spent a few days with Masey in Berlin to plan the exhibition and visit the city. Masey "spoke no German and yet managed to dominate all conversations by decibel count and similar means. But, for some reason that remains a total mystery to me as well as to him, he and I became fast and utterly devoted friends." Indeed, they remained so, working together on a number of projects for USIA, until Peter Blake's death in 2006.

“We live behind Communist guns in this encircled city. That is a geographical fact. But we belong to the West and we will continue belonging to the West.”

Willy Brandt, Mayor, West Berlin, December 7, 1958

Cloudy

KALAMAZOO GAZETTE

HOME EDITION

125TH YEAR—NO. 154. 52 PAGES SECTION 1—PAGES 1-12. KALAMAZOO, MICHIGAN, SUNDAY, MARCH 23, 1958. Phone FI 5-3511 ★ PRICE

USIA WORKERS, MRS. ELEANOR DULLES, DISCUSS KALAMAZOO EXHIBIT FOR BERLIN
From Left: Jack Masey, Max R. Grossman, USIA: Will Burtin, Display's Designer

Jack Masey (left) with Eleanor Dulles and Will Burton (right) in Kalamazoo.

Visitors arriving at the front of the Marshall House for the Medicine USA exhibition.

Medizin-USA

A display of architectural models and photographs.

A large aerial photograph of New York showed the scale of urban building in the USA.

AMERICA BUILDS

In designing the "America Builds" exhibition, Peter Blake and his partner Julian Neski sought to provide Marshall House visitors with a sense of the scale of American urban buildings. Blake also said that because the Marshall House was so "ungainly… it became clear to me that the only way to tackle the job was to use the exhibition to transform its interiors into a series of very different spaces." In other words, one of the roles of the design of the exhibition itself would be to conceal the space it was in.

Photographs combined
with models in the display.

> Photographic details of notable
American buildings.

The "America Builds" exhibition featured full-scale, impeccably detailed mock-ups of the facades of some of the most noteworthy modern skyscrapers in the United States. To give visitors a realistic impression of the enormous height of these buildings, as well as to emphasize their transparency and grace, Blake and Neski made use of special lighting and orthogonally reflective mirror devices.

Among the skyscrapers featured were three in New York City: the Union Carbide Building by Skidmore Owings & Merrill, the Lever House by Gordon Bundshaft, and the Seagram Building by Mies van der Rohe. Another, located in Pittsburgh, was the Alcoa Building by Harrison and Abramovitz. One of the most popular exhibits in the Marshall House was a 360° panoramic photograph installed in a huge circular room, which showed a glowing Manhattan skyline at night taken from the McGraw-Hill Building.

The "America Builds" exhibition was up against some considerable competition: as part of the Interbau event, over twenty architects from Germany and around the world, including Walter Gropius, Alvar Aalto, Oscar Niemeyer and Sven Markelius, had been invited to construct new buildings for the city, mainly in the Hansa area, though Le Corbusier insisted on a site for his Unité d'Habitation apartment building away from the others. Another important building was the Kongresshalle, designed by the American architect Hugh Stubbins and funded by the Benjamin Franklin Foundation. (This structure, nicknamed "the Pregnant Oyster," used concrete arches to hold a suspended roof; it collapsed twenty years later, an event predicted at the time by the architect and theorist Frei Otto.)

The Interbau project and the exhibition at the Marshall House were not just intended for the delight of West Berliners. The target was as much the East, both East Berlin and the Eastern bloc: for a number of years travel from the Eastern zone to the Western was unregulated, and so the exhibitions and events could be seen by and influence East Germans as well; there were even different ticketing systems for East and West Berliners so that the number of East German visitors could be monitored. Although many structures were Modernist in style, the American buildings shown in Blake's exhibition and those constructed for Interbau were a deliberate and provocative contrast to the centrally controlled and ideologically dominated work being done in the eastern part of the city. The irony was of course that East Berlin was a capital city while West Berlin was not.

The very complexity of Berlin's situation made the city an exciting and challenging place to be. "I think, in retrospect," Blake wrote later of the time he spent there with Masey, "that we behaved like a couple of wild besotted kids, night after night. We had the time of our lives. No city that he or I knew, in those years, could compare with Berlin in sheer, unbridled madness; and we loved every moment of it."

A key figure in Berlin at the time was Eleanor Dulles, the sister of John Foster Dulles, the Secretary of State, and Allen Dulles, head of the CIA. Dulles was a Harvard-trained economist who, having organized the State Department's "Berlin Desk" in 1949, was involved in its post-war economic planning. Eventually known as "the Mother of Berlin" for her unstinting efforts to restore that capital's economy and culture after the devastation of World War II, she was on the board of the Benjamin Franklin Foundation, where she was instrumental in planning the Berlin Kongresshalle and later, the Berlin Medical Center. At her death in 1996, aged 101, the State Department issued a press release that noted, "... her accomplishments in the Department of State ... were the result of her own outstanding abilities and effort. Like other women in government service in mid-century, she had to confront persistent sex discrimination, prejudice against hiring and advancement of women, and obstacles to obtaining important assignments."

Visitors and photo enlargements of architects at work.

MEDICINE-USA

The Medicine-USA exhibition, displayed at the Marshall House during the 1959 Berlin Industries Fair, was designed by Peter G. Harnden. It was part of a long-term medical exchange program between West Berlin and the USA, which was to lead to the funding and construction of a medical center in Berlin.

The overall theme of the Medicine-USA exhibition was the progress made in recent years in all branches of medical science in the US. Depicted in the exhibition was the role played by labor, industry, government and private foundations in providing medical education, research and care.

Exhibits in the Medicine-USA show featured a fully-equipped, fully-functioning modern operating theater, a full-sized "recovery room" for post-operative care, a neo-natal nursery, a massive nine-ton Cobalt Therapy Unit, an electronically-controlled set of "magic-hands" used to manipulate radioactive materials, and a display showing how isotopes were used in medical research. Live demonstrations, special lectures and film showings, all suitable for general audiences, enlivened the show. Demonstrations were continued outside the Marshall House, where American Boy Scouts re-enacted first-aid techniques near a display of air-rescue equipment that featured a US Army helicopter.

Medicine-USA was seen by more than a half-million visitors in West Berlin in the fifteen days of its showing (September 12th to 25th). After Berlin, the exhibition was redesigned for showings in Dusseldorf, Hamburg and Munich.

The replica operating theater with the famous DeBaky "heart/lung" machine attracted many visitors.

The neo-natal clinic was also a popular display.

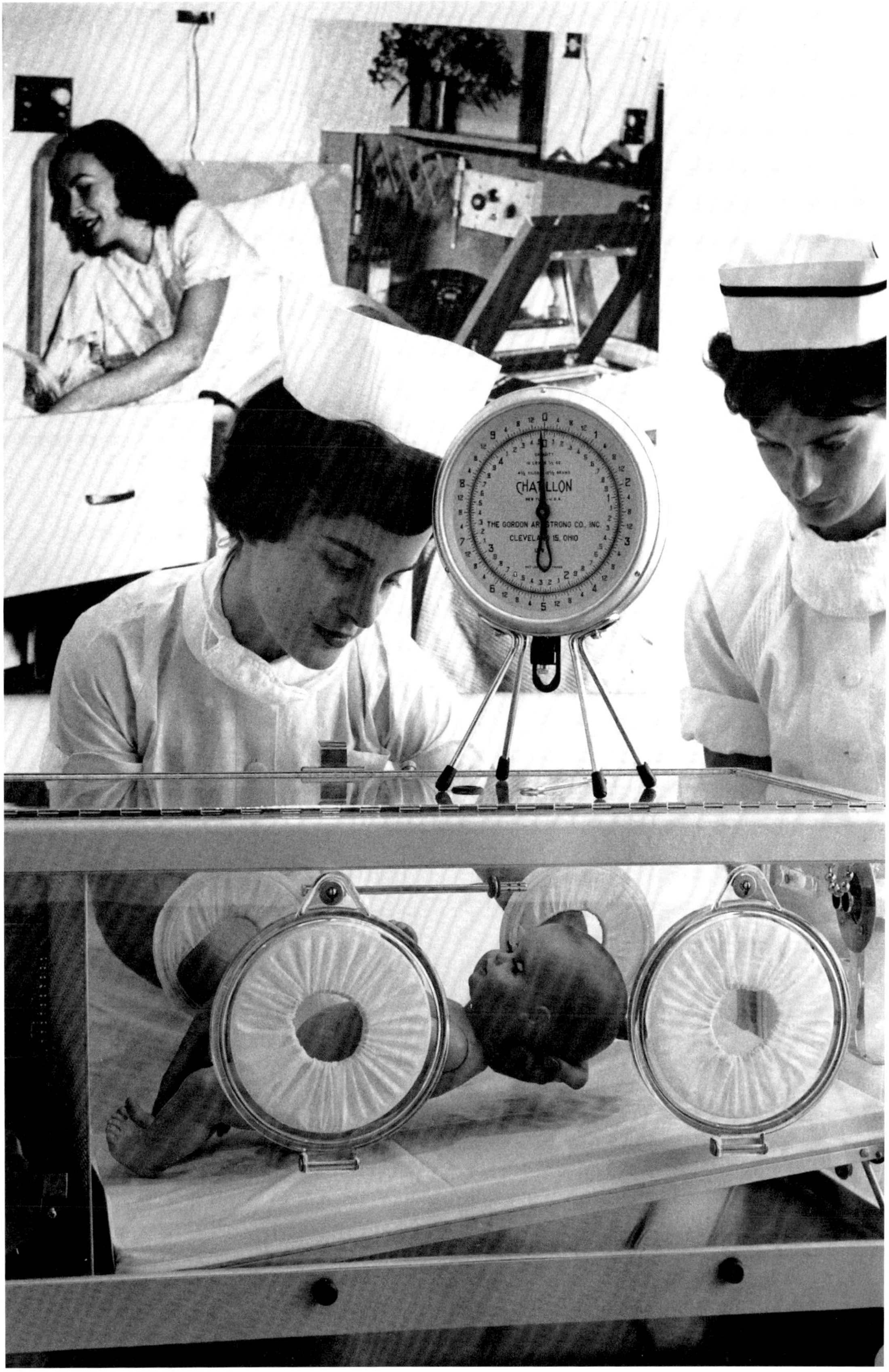

An air-sea rescue helicopter was part of the outdoor display.

American Boy Scouts demonstrating first aid.

The Magic Hands apparatus for handling radioactive isotopes.

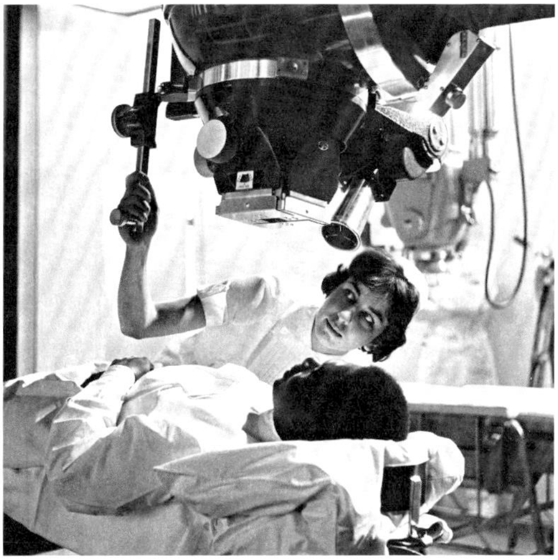

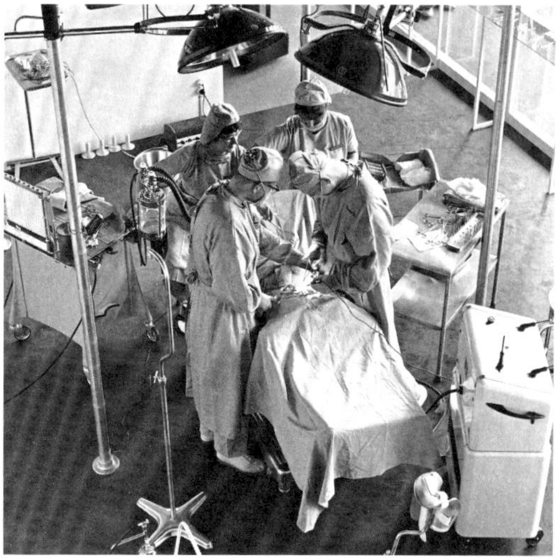

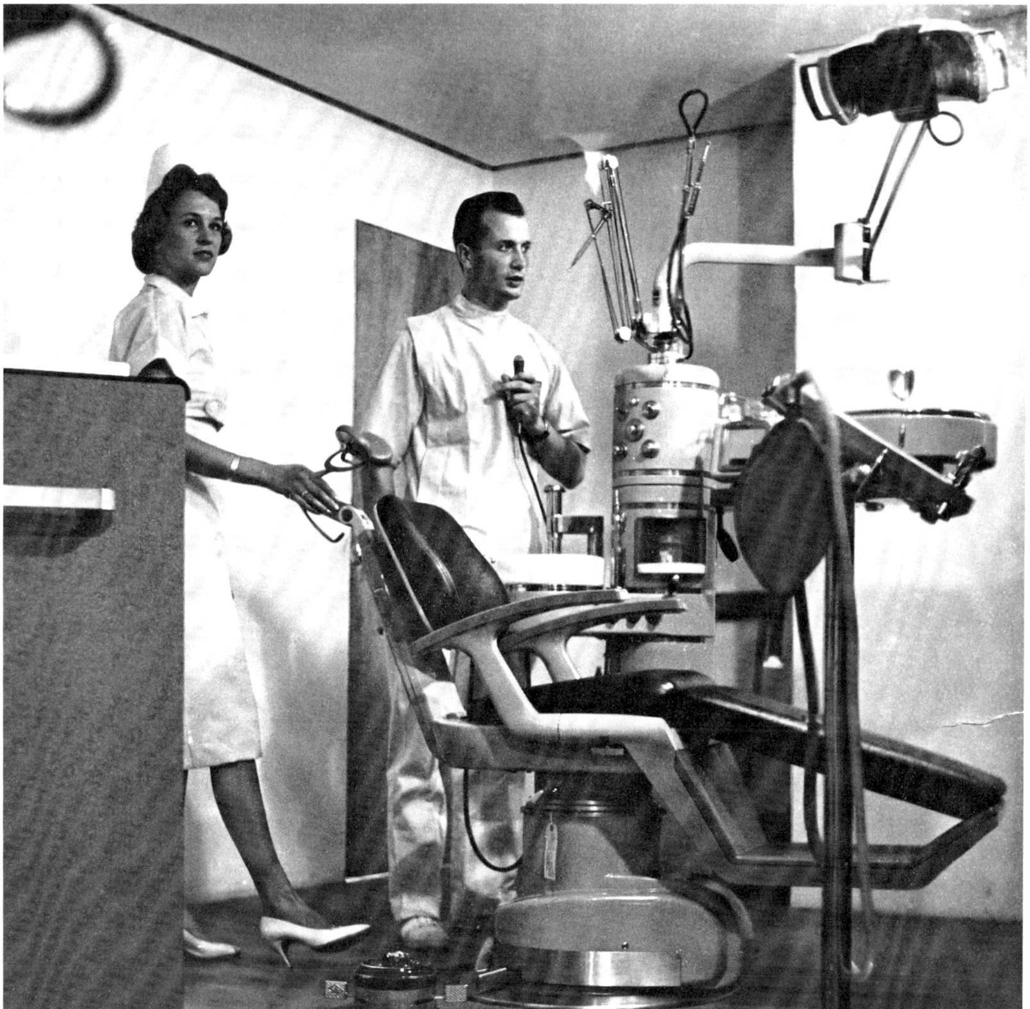

The cobalt radiotherapy machine
and a modern dentist's surgery.

KALAMAZOO ... AND HOW IT GREW

In 1958 the Marshall House housed an exhibition about a typical midwestern city, Kalamazoo, selected both for the whimsy of its name and as an example of postwar changes in American living. The exhibition was designed by a former Berliner, Will Burtin, and concentrated on the everyday life of Kalamazoo's citizens, and the main industries of the city: paper manufacture and the Gibson guitar factory. Both these elements figured prominently in the exhibit design. The show was visited by tens of thousands of Berliners during its two-week run.

Kalamazoo was chosen as a subject for Western Europe (a smaller version had earlier been shown in London) because it most appropriately illustrated a changing American community of many races and nations. Burtin's design brought together—through the metaphor of a roll of paper—the lives of everyday people and their relationship to their city and its industries in a way many could understand.

Burtin, born in Germany in 1908, had attracted broad national attention for his design work in Germany in the 1930s. Eventually approached by Hitler to design for the Nazis, he would later recall, "They were insidious and I couldn't refuse and still stay in Germany." In 1938, Burtin and his wife, Hilde, fled to the United States where he set up a successful design practice in New York City. The choice of designers such as Blake and Harnden who had fought against the Nazis, and Burtin who had refused to co-operate with them, is witness to the sensitivity of Berlin as a Cold War front line.

Photographs of the everyday lives of 20 citizens of Kalamazoo formed the opening part of the exhibit and a continuing theme.

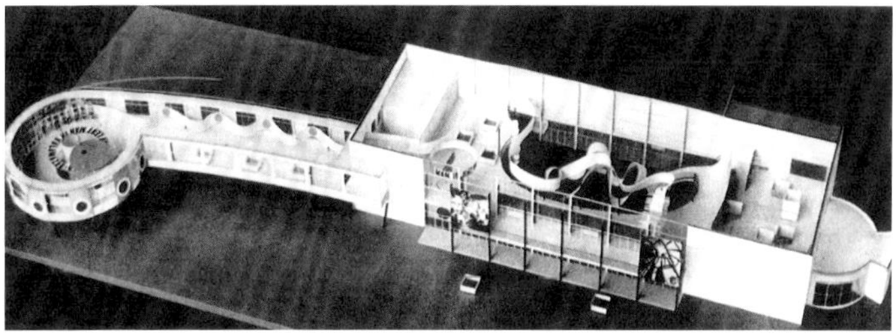

Will Burtin outside the Marshall House.

Model of the exhibition.

Views of the exhibition including the roll of paper that formed a central theme.

> General view of the exhibit with the giant Gibson guitar.

ATOMIC EUROPE

UNITED STATES PAVILION
UNIVERSAL AND
INTERNATIONAL EXPOSITION
BRUSSELS, BELGIUM
1958

"A World's Fair is an art form, a combination of beauty and bombast, and an expression of a complex idea involving the arts, national and individual prestige, uplift and the universal hankering for a holiday."

George R. Leighton's description has applied to all world's fairs since the success of the Great Exhibition held in John Paxton's Crystal Palace in London in 1851. World's fairs proliferated in such numbers that, by the beginning of the twentieth century, competing world's fairs were often held simultaneously. To regulate the frequency and quality of what had become hundreds of competing international exhibitions, the Bureau of International Exhibitions (the BIE, as it would come to be known) was created in Paris in 1928. The BIE established two types of world's fair categories: Category One Registered Exhibitions, in which nations would be invited to build their own pavilions, and could not exceed six months in duration. Category Two Recognized Exhibitions, in which nations were not obliged to erect their own pavilions, nor could they exceed three months in duration. Both categories came to be known as Expos.

The more prestigious Category One Expos are considered the "Olympics" of world's fairs; like the Super Bowl and the World Cup, these major Expos attract the interest of millions around the world because they offer a dazzling display of competition between national teams. At Expos, however, the teams are not playing games: rather, they competed by presenting to the world examples of a nation's best architecture, technology, arts, crafts, manufacturing, and performing arts. And in so doing, they sometimes, somehow, change the world.

< Visitors inside the American Pavilion.

The Atomium, symbol of the Brussels Exposition.

As George Nelson noted in 1953: "It is an odd thing, but true, that when one begins to trace the developments in architecture, structure, interior design and related areas, the old expositions turn out to be remarkably accurate guides to doing things. Paxton's Crystal Palace was a prefabricated structure entirely done in metal and glass, and its implications are not fully exhausted a century later."

The temporary nature of Expos amplified the ability of architects and designers to explore new ideas and technologies. Nelson understood well the value of this opportunity: "In a temporary exposition … the need to be practical does not exist … The result can be fun. It is surprising how often it can be significant fun."

The first Category One Expo sanctioned by the BIE after World War II was the Universal and International Exposition in Brussels in 1958. The next Category One Expo would be held in Montreal in 1967, followed three years later by a third, in Osaka, Japan. US participation at these three Expos is covered in this book. Expo '58 represented for the host country and for Europe generally an opportunity to demonstrate the reconstruction and economic development that had taken place in the decade since the Marshall Plan. The theme of progress towards peace was symbolized in the Atomium structure that dominated the fairground—and which remains in place today. The atomic metaphor was a prevalent one in the design of the early 1950s, used in fabrics, furniture and product design. It would be easy with hindsight to read such design as ironic, especially in Europe, which would again have become a battleground should the superpowers of the USA and the USSR have resorted to resolving their differences by force of arms. But as Spencer R. Weart has pointed out, attitudes in the immediate aftermath of Hiroshima and Nagasaki towards atomic power and the atomic bomb were ambivalent, as the long-term consequences of the use of atomic weapons was still unclear.

By the mid-1950s and with the increasing deployment of hydrogen and atom bombs, the nature of the nuclear risk had become apparent: the year 1958, for example, saw the formation of the Campaign for Nuclear Disarmament in the UK. Europe was also another kind of battleground, on a political and intellectual level. The CIA was covertly funding a number of organizations to promote democratic values and stem Communist influence, such as the Council for Free Europe; in France and Italy, there were well-established Communist parties, both of which still retained public kudos for their resistance role in the Second World War. European opinion was also considered as influential on the "non-aligned" countries, even if many were former colonies of European states. Making a good impression in Europe was thus seen as being of crucial importance to the US government, and for the additional reason that the pavilion would also be seen by a considerable number of American visitors, both tourists and US servicemen based in Europe. In addition, what the US did and how the nation was presented at Brussels would come under considerably more public, press and Congressional scrutiny, and in a context, unlike Berlin, where visitors could compare the American effort with that of the Soviets and other countries.

Why the State Department rather than the USIA or the Department of Commerce took on the task is not clear: Haddow argues that the official invitation included a cultural element, which would have ruled out the Department of Commerce, but hardly the USIA. What is known is that USIA's funds had been severely cut by Congressional opponents of its director, Arthur Larson.

The official invitation to participate, which was sent in June 1954, went, as a matter of protocol, directly to the State Department. The Department seems to have moved quite slowly on the invitation as it was not until 1956 that they appointed an architect, Edward Durrell Stone, and only a few months later, in October, appointed Howard Cullmann as Commissioner General. Cullmann set about building a team for the task among his contacts: one of the most important of these was James Plaut, director of the Boston Institute of Contemporary Art, who became Deputy Commissioner in November, and who would be responsible for supervising the design and much of the detailed content.

This sudden, late rush to make decisions may well be linked to a telegram sent on July 24, 1956 by the State Department to the American Embassy in Brussels. Marked "Confidential," it begins: "Extremely disturbed reported Belgian decision Chinese Communist invitation Brussels Fair. Inform Spaak our view invitation would offer major propaganda opportunity, which Chinese Communists would exploit fully." While the telegram does not give a source for the report, it is clearly instructing the Ambassador to express America's concern to Paul Henri Spaak, the well-respected Belgian Foreign Minister, and goes on to threaten the possibility of American withdrawal if the Chinese were to take part, as well as denying that Chinese participation in various European trade fairs forms any sort of precedent, and citing a recent Congressional vote against Communist China joining the United Nations or its agencies. The Belgian authorities later denied an invitation had been issued, but the possibility of the participation of the brutal and aggressive Chinese regime (to paraphrase the telegram) seems to have awakened State to the urgency of decisions about the project.

Threat of Chinese presence or not, the stage was set for a major confrontation, given the fact that the Soviets were planning to build the largest and most expensive pavilion at the Expo. For them as well, the Communist history of industrial Northern France, as well as that of Belgium and Holland, was a good opportunity for a strong show of solidarity.

The American Pavilion.

DESIGN AND DISAGREEMENT

During the conceptual design phase of any major national pavilion, there are inevitable differences which surface between the pavilion's management, the architect and the exhibit designer. It was no different at Expo '58. The opinions of Howard Cullman, the Commissioner General of the US Pavilion, about exhibit design differed from those of its architect, Edward Durrell Stone. And both differed with Peter G. Harnden and Bernard Rudofsky, who were responsible for the design of the pavilion's interior and its exhibits.

Stone had been working on the exhibits as well as the building for several months before Cullmann was appointed, but it was not until the following April, hardly more than a year before the Fair was to open, that Stone was told he was not to be responsible for the exhibit design. Cullmann's office had begun by interviewing prominent American citizens, including bankers, diplomats and academics, for their views about how America should be represented in Brussels. They even went to see Walt Disney (the first Disneyland had opened in 1955) but reported that he "came up with almost nothing." Walter Paepcke, more positively, advocated putting American culture rather than material prosperity at the forefront. The former US Ambassador to Denmark, Mrs. Eugenie Anderson, offered a very detailed proposal, stressing both technological strengths and cultural achievements, while being honest about America's weaknesses "since self-criticism is one of our greatest strengths." She also emphasized European ignorance of America's history and way of life.

This process of investigation soon led the team to MIT, and in particular to their CENIS group, where for some time the prominent economist Walt W. Rostow had been advocating a more frank and determined approach in American public diplomacy. This connection led in April 1957 to a weekend conference that brought together some thirty academics and other specialists urgently charged with creating a brief for the exhibition. Sketches and notes in the MIT archive suggest some rather simplistic solutions: one, headed "Man in America," proposes a circular relief map of the USA in the center with concepts radiating out from it in an improbable sequence—one segment reads "leisure, health, God, food, rent, unemployment" in an outer ring, with annotations for an inner ring that end, between unemployment and the relief map, with the word "problems." Another sheet of notes identifies five themes: Land and People, Science and Technology, Culture, Life and Work, and American Values at Work. The committee's report endorsed these five topics, and submitted an additional report.

This referred to the “unfinished business” that America needed to address at the Exposition, in particular the issues of soil erosion, urban decay and race relations. This report takes its cue from Anderson’s proposal: unemployment and the “Negro problem” also figure in the handwritten notes. This suggestion, reinforced as it was by Mrs. Anderson’s call for self-criticism, was to have important consequences.

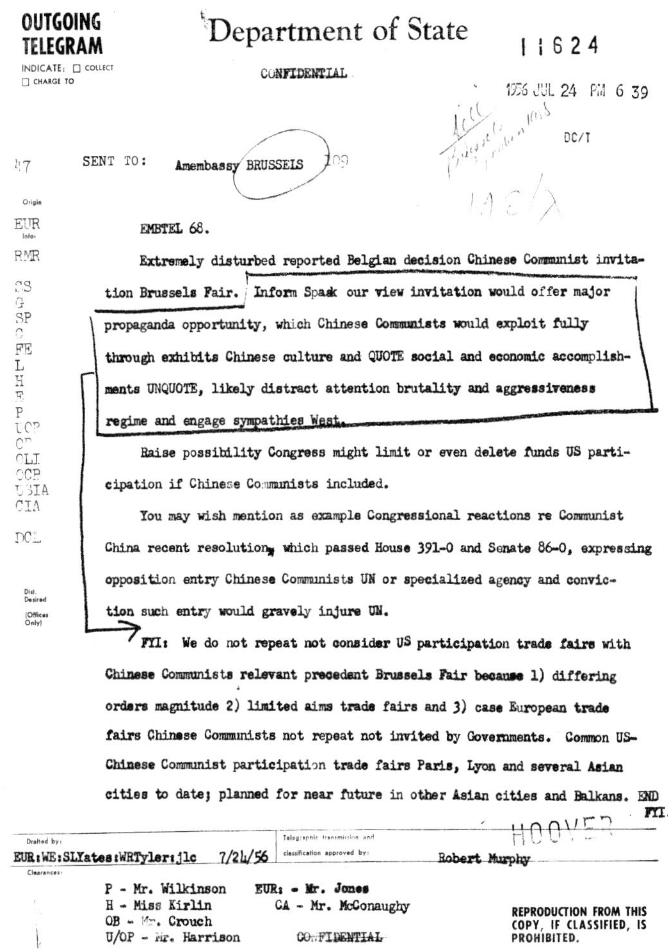

OUTGOING TELEGRAM

INDICATE: ☐ COLLECT ☐ CHARGE TO

Department of State

11624

CONFIDENTIAL

1956 JUL 24 PM 6 39

DC/T

SENT TO: Amembassy BRUSSELS 109

Origin EUR

EMBTEL 68.

Extremely disturbed reported Belgian decision Chinese Communist invitation Brussels Fair. Inform Spaak our view invitation would offer major propaganda opportunity, which Chinese Communists would exploit fully through exhibits Chinese culture and QUOTE social and economic accomplishments UNQUOTE, likely distract attention brutality and aggressiveness regime and engage sympathies West.

Raise possibility Congress might limit or even delete funds US participation if Chinese Communists included.

You may wish mention as example Congressional reactions re Communist China recent resolution, which passed House 391-0 and Senate 86-0, expressing opposition entry Chinese Communists UN or specialized agency and conviction such entry would gravely injure UN.

FYI: We do not repeat not consider US participation trade fairs with Chinese Communists relevant precedent Brussels Fair because 1) differing orders magnitude 2) limited aims trade fairs and 3) case European trade fairs Chinese Communists not repeat not invited by Governments. Common US-Chinese Communist participation trade fairs Paris, Lyon and several Asian cities to date; planned for near future in other Asian cities and Balkans. END FYI.

HOOVER

Drafted by: EUR:WE:SLYates:WRTyler:jlc 7/24/56

Telegraphic transmission and classification approved by: Robert Murphy

Clearances:

P - Mr. Wilkinson
H - Miss Kirlin
OB - Mr. Crouch
U/OP - Mr. Harrison

EUR: - Mr. Jones
CA - Mr. McConaughy

CONFIDENTIAL

REPRODUCTION FROM THIS COPY, IF CLASSIFIED, IS PROHIBITED.

The State Department message about the threat of China’s involvement in the Fair.

> Interior of the completed Pavilion.

"Brilliantly heralded by flags, fountains and apple blossoms, architect Edward Stone's magnificent drum is in a unique position to steal the show."

The *Architectural Forum,* June 1958

THE ARCHITECT'S PLAN

Edward Durrell Stone was an architect who has since come in for considerable vilification, especially from Tom Wolfe in his book *From Bauhaus to Our House,* for not being Modernist enough. At the time of his appointment he had already been commissioned by the State Department as architect for the American Embassy in New Delhi, a project that was generally well received for its blending of a contemporary style with sensitivity to the traditions of Indian art and architecture. It is likely that Stone's name was among those the American Institute of Architects suggested to the Foreign Building Office in the State Department as a suitable architect for the design of an important American Embassy.

The six and a half acre site assigned to the US Pavilion was a triangular one, with access on all three sides. The only stipulation made by the authorities was that three willow trees growing on the site be left undisturbed. Set in the center of the Fair's grounds, the US Pavilion would be between those of the USSR and the Vatican: comments about Heaven, Purgatory and Hell were soon being tossed around! Stone's proposed solution was elegant and innovative. He designed a circular building, in which the roof would be entirely supported by the walls, so leaving the interior space uncluttered by columns or other support structures. The roof would be of plastic, at the time a relatively new architectural material, supported by steel tension cables. Underneath the oculus, the circular opening at the center of the dome's roof, there would be a dramatic light tube illuminating an abstract wire-frame sculpture by Richard Lippold, set in a pool and surrounded by fountains. A similar external pool and fountains, with a giant mobile by Alexander Calder, would face one of the main entrances. This entrance, not surprisingly, was

Plan of the Exposition site.

Postage stamp celebrating the American Pavilion.

Visitors arriving at the American Pavilion.

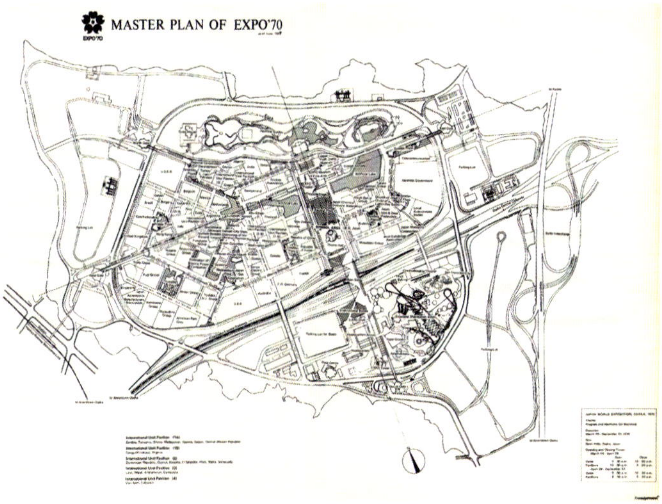
MASTER PLAN OF EXPO'70
EXPO'70

UNIVERSAL AND INTERNATIONAL EXHIBITION·1958
BRUSSELS
UNITED STATES
PAVILION
U.S.POSTAGE 3¢

directly across from the USSR's pavilion. The interior space would be relatively uncluttered: a celebration of culture, not a vehicle for information.

Cullmann and his colleagues, however, realized that what visitors would want was not an ambassadorial foyer but a range of facts, images and objects that would convey a panoramic view of the American experience. They provided the reports of the MIT meeting, along with the recommendations of advisory committees, to Peter G. Harnden, a former Marshall Plan designer now running an independent design office in Paris, with close associations with the US government. He rapidly realized that the sheer volume of material to be presented would require a two-level interior space, and would need a dramatic gesture to link the two. Plaut and Harnden's solution was a runway from the upper level over the central pool (bereft of its fountains, to Stone's chagrin), for a fashion display with live models that would be one of the main events of the pavilion. Plaut and Harnden asked *Vogue* to help with the fashion show.

Catwalk for the fashion show in the center of the Pavilion.

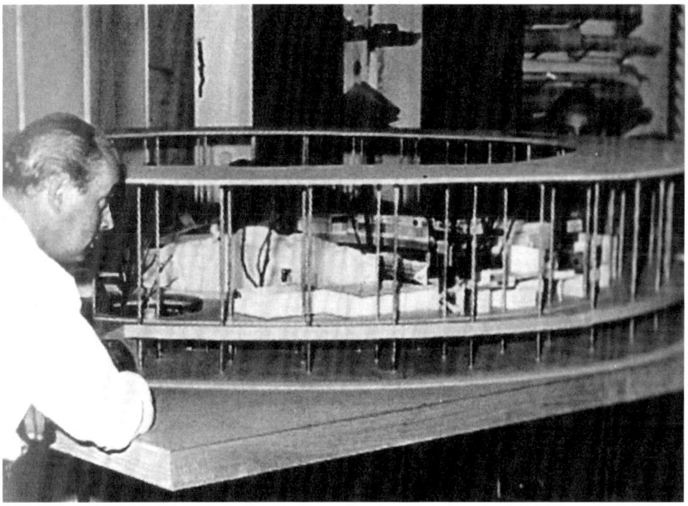

Peter Harnden studying a model of the Pavilion.

The interior seen from the upper level.

THE CONTENTS

The planners of the US Pavilion did not completely overlook the Atomium theme of the Brussels Expo: a nuclear energy exhibit was included. But the planners were eager to show their European audience—witnesses of American military prowess in World War II—that the US had a human side too.

The fashion show, which was organized by Lee Canfield, the sister of Jacqueline Kennedy, was, quite literally, given center stage on a long ramp in the middle of the US Pavilion. Other examples of the softer side of American life were also represented. A giant mural by Saul Steinberg presented a parody of American stereotypes. A Circarama film show, a display of voting machines (a technology unfamiliar to Europeans) and the "Streetscape" exhibit on American urban life were popular with visitors. (The success of the "Streetscape," by the newly formed New York design firm of Chermayeff and Geismar, would lead to future collaborations between the young designers and the US government.)

A guide demonstrating the IBM RAMAC computer.

Children at play in the children's exhibit.

While the US exhibition was generally well received by European visitors and press, the abstract paintings in the American art show were a source of controversy. A reviewer in a Belgian trade journal noted in June 1958, of having seen "a number of average visitors confronted with huge canvasses showing only a few streaks of color, who became sincerely concerned about the sanity of the American people."

While the reviewer doubted the popularity of American abstract art with most visitors to Expo '58, his summation of the US exhibition was generally positive. "Everybody knows that America is a giant. The planners of your exhibit have shown that, not withstanding his size, this giant is a warm human being ... not constantly boasting about being bigger and stronger and richer than everybody else ... America (has shown) itself ... as an essentially humane country."

Visitors in the "Streetscape" exhibit designed by Chermayeff and Geismar.

YIELD
RIGHT
OF
WAY
DEPT OF TRAFFIC
CURB YOUR DOG
VIOLATION
$25.
FINE
STOP
LOUIS

A model on the catwalk in the fashion show.

Visitors watching the fashion show.

Two views of the mural created by Saul Steinberg for the American Pavilion.

UNFINISHED BUSINESS

As planning within the State Department for the pavilion at the Brussels Expo was being developed in 1956 and 1957, events were playing out in the US and abroad that would result in the creation of one of the most successful examples of architectural propaganda ever undertaken by the United States abroad.

The planners had much to be concerned about: In 1956, when the US complained bitterly at the UN about the brutal Soviet repression of the Hungarian uprising, the Soviets countered by pointing to the US's deplorable record with race relations: the brutal murder of Emmett Till in Mississippi and Rosa Parks's arrest in Alabama was soon followed, in 1957, by the school desegregation crisis in Little Rock, Arkansas. President Eisenhower was forced to send National Guard troops to Little Rock to protect nine black students who had attempted to enroll in an all-white high school from angry white mobs. News images and reports of the Little Rock incident were broadcast worldwide. At the UN,

Visitors in front of the
"Unfinished Business" structure.

"It is a ridiculous and unpardonable mistake [for the U.S. State Department] to be part and parcel of any effort, by any American, to 'Foul our Nest'."

L. Mendel Rivers, Member of Congress

the Russians accused the US of "racial terror" in Arkansas, using the incident to counter US charges of Soviet repression. The Soviet complaint hit home in the MIT planning group. It was a period of heightened racial tensions at home and declining US prestige abroad: Walter W. Rostow warned the planning group that the desegregation issue in the US would be "underlined rather than evaded by omission." The MIT Group decided to face Soviet criticisms head-on in Brussels by proposing the creation of an exhibit specially designed to present America's domestic problems forthrightly as part of an ongoing struggle.

The "Unfinished Business" exhibit, created by *Fortune* magazine (part of the Time-Life group owned by Henry Luce) with the State Department, was originally conceived as a way to face—openly and directly—anticipated negative Soviet propaganda about domestic problems within the US. As a piece of formal presentation "Unfinished Business" can hardly be faulted: it is honest at the outset, winning the visitor's confidence through openness in the first section, documenting efforts towards progress for the underprivileged in the second, while the third simply presents images of the hope for the future. The whole exhibit acts as a timeline: past, present, future. The overall message was that the business might be unfinished, but progress was being made.

In a highly innovative way the content of the interior was also to be conveyed through the exterior of the pavilion: the first section—a "chaotic crystal"—had a distressed and crumpled exterior, its facets painted in black, brown and white. The center section was a much

> Visitors entering the first section of "Unfinished Business."

The unfinished work
Le travail inachevé
Het onbeëindigde werk

simpler polyhedron, its larger facets painted in bright colors, and the third and final section was optimistically smooth and bright, the exterior painted in large panels of blue and white. Few other exhibits at that time would have tried to act so allegorically, and seeing images of the exhibit today its approach is still remarkable, even though it must be granted that the visitor would only understand the symbolism of the exterior after a visit to the interior. Because of the sensitivity of the issues, visitors would be taken through the exhibit by guides who would explain the contents and answer questions—feedback from visitors suggested that this was the right approach. As an exhibition concept it is unusual and strong: the mirroring of inner and outer messages must have been unusual and intriguing, especially as the message was reinforced by the color scheme. Most European visitors, surveys showed, felt that the exhibition had deepened their understanding of America's problems and its attempts to deal with them: a positive result. American visitors, on the other hand, had very different views.

Model of the three sections that make up "Unfinished Business."

The second and third sections.

DEATH TOLL 72 IN EAST'S
Gallagher Gets Life Term For Killing 2
YOUNG GOP
Disease Fears Rise in Suburb With No Water
Ike Calls Dulles on Mideast
WINDS WAFT SMOG OVER THE HUDSON
to Halt Water Pollution
Serious Health Problem
STAGGERING LOSS FEARED IN
IRISH 'RAID' JUST A HOAX
The Det

Is Ordered Against Faubus
Governor Says A Threat On Life Made In North
It's YOUR
Housing in Newark
Segregation
News

As news of the exhibit leaked out before its opening, Southern members of Congress mounted a fierce effort to block it from opening at all. Senator Herman Talmadge of Georgia complained to Secretary of State Dulles that the exhibit represented "an unwarranted invasion of the rights and prerogatives of the states of the South" where "segregated society has proved to be in the best interests of all races concerned." Failing to prevent the opening of the exhibit, Talmadge, along with Senator Strom Thurmond and Representative Mendel Rivers mounted heavy pressure on the Eisenhower Administration to close the Exhibition. After a flurry of urgent telegrams between Washington and the Embassy in Brussels about the exhibit's contents, on the evening of May 8, 1958 (the same day that a highly favorable review was published in the Brussels press), acting Secretary of State Christian Herter ordered the exhibit to be closed to visitors. Three weeks later, modest changes eliminating "all features heretofore considered objectionable" were made to the exhibit: among them, a sign pasted on a version of a controversial photo of mixed race children at play stated that the image did not represent US policy but rather the individual freedom of choice available in the US. The exhibit was reopened on a test basis for small groups whose visits were now "controlled by trained guides."

The interior of the first section with newspaper cuttings on the walls.

Exterior of the first section.

The resulting exhibit consisted of a separate structure alongside the main pavilion, divided into three sections, each about 20 foot long, and 12 high and wide (6 × 4 × 4 meters). The exhibit was sponsored by *Fortune* magazine and designed by Leo Lionni, then art director of *Fortune* and later to make a reputation as an author of children's books. He realized that the need to create an independent structure was also a design opportunity. The exterior as well as the interior could be used to reinforce the central message of the exhibit. The agenda for the exhibit was derived from the Rostow report, as is made clear from the account of the design process, written at the time by Judith Murphy who was part of the design team. According to Murphy, because of the limited space for the exhibit, Lionni decided to concentrate on "three key areas: the Negro, the City and Nature."

Exterior of the second and third sections.

The interior of the second "progress" section.

US Embassy reviews of European press reactions at the time indicated that Lionni's exhibit had largely succeeded in communicating what its planners had originally hoped. The Belgian newspaper *Le Peuple* wrote, "Let's face it: only strong democracies are in a position to talk as well of their qualities as their faults. Our thanks to the United States for having demonstrated this at the Heysel." But, as successful as Lionni's extraordinary design may have been in the view of its European visitors, in Washington it was doomed, a victim of its own success. By mid-August 1958, the State Department had redefined the nature of America's "Unfinished Business": a new exhibit on Public Health was installed amid protests from the US guides and a call for a Congressional investigation into the closing of "the most effective propaganda in the American pavilion." It was an unworthy end to one of the most successful examples of architectural propaganda ever attempted by the United States.

"[The Unfinished Business exhibit could be] the bombshell of the U.S. [exhibit] as well as the section of the show most likely to blow up in our planners' faces as visiting American may not see eye-to-eye on all the answers."

The Saturday Evening Post, January 25, 1958

The exterior of the third section.

The interior of the third section with images of rice-farming and urban redevelopment.

> The closing image of children dancing in a field.

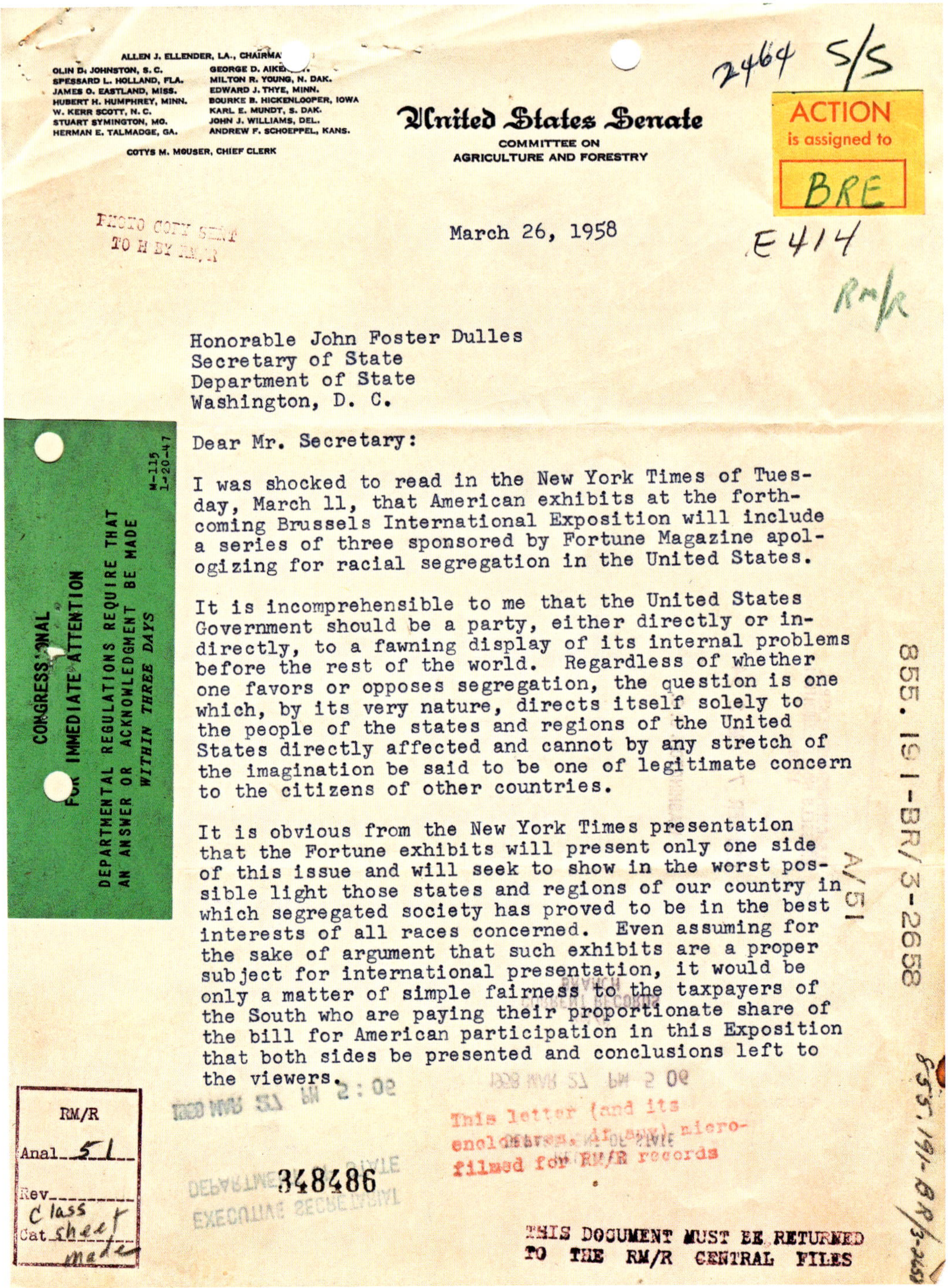

ALLEN J. ELLENDER, LA., CHAIRMAN
OLIN D. JOHNSTON, S. C.
SPESSARD L. HOLLAND, FLA.
JAMES O. EASTLAND, MISS.
HUBERT H. HUMPHREY, MINN.
W. KERR SCOTT, N. C.
STUART SYMINGTON, MO.
HERMAN E. TALMADGE, GA.
GEORGE D. AIKEN, VT.
MILTON R. YOUNG, N. DAK.
EDWARD J. THYE, MINN.
BOURKE B. HICKENLOOPER, IOWA
KARL E. MUNDT, S. DAK.
JOHN J. WILLIAMS, DEL.
ANDREW F. SCHOEPPEL, KANS.
COTYS M. MOUSER, CHIEF CLERK

United States Senate
COMMITTEE ON
AGRICULTURE AND FORESTRY

2464 S/S

ACTION is assigned to BRE

PHOTO COPY SENT TO H BY RM/R

March 26, 1958

E414

RM/R

Honorable John Foster Dulles
Secretary of State
Department of State
Washington, D. C.

CONGRESSIONAL FOR IMMEDIATE ATTENTION
DEPARTMENTAL REGULATIONS REQUIRE THAT AN ANSWER OR ACKNOWLEDGMENT BE MADE *WITHIN THREE DAYS*
M-115 1-20-47

Dear Mr. Secretary:

I was shocked to read in the New York Times of Tuesday, March 11, that American exhibits at the forthcoming Brussels International Exposition will include a series of three sponsored by Fortune Magazine apologizing for racial segregation in the United States.

It is incomprehensible to me that the United States Government should be a party, either directly or indirectly, to a fawning display of its internal problems before the rest of the world. Regardless of whether one favors or opposes segregation, the question is one which, by its very nature, directs itself solely to the people of the states and regions of the United States directly affected and cannot by any stretch of the imagination be said to be one of legitimate concern to the citizens of other countries.

It is obvious from the New York Times presentation that the Fortune exhibits will present only one side of this issue and will seek to show in the worst possible light those states and regions of our country in which segregated society has proved to be in the best interests of all races concerned. Even assuming for the sake of argument that such exhibits are a proper subject for international presentation, it would be only a matter of simple fairness to the taxpayers of the South who are paying their proportionate share of the bill for American participation in this Exposition that both sides be presented and conclusions left to the viewers.

855.191-BR/3-2658

A/51

RM/R
Anal 51
Rev
Cat class sheet made

1958 MAR 27 PM 2:06

348486

DEPARTMENT OF STATE
EXECUTIVE SECRETARIAT

This letter (and its enclosures, if any) microfilmed for RM/R records

THIS DOCUMENT MUST BE RETURNED TO THE RM/R CENTRAL FILES

Letter from Senator Talmadge to the Secretary of State protesting the planned exhibition.

Honorable John Foster Dulles page 2

The participation of the Department of State in arrangements for such exhibits constitutes an unwarranted invasion of the rights and prerogatives of the states of the South and constitutes a gross insult to the good name and conscientious citizens of that region. It amounts to a repudiation by the Department of State of its constitutional responsibility to represent before the world the best interests of all Americans regardless of their beliefs, persuasion or place of residence.

I wish vigorously to protest the role of the Department of State in the preparation and sponsorship of these exhibits and the expenditure of American tax funds toward that end. I call on you, in the name of the people of Georgia whom I have the honor to represent in the Senate of the United States, to repudiate the impression thereby given that the Department of State has set itself up as an arbiter of right and wrong in matters of purely internal import.

A report from you as to the extent of American participation, from the standpoints of both finances and personnel, in the planning, preparation and installation of these exhibits at the Brussels International Exposition will be deeply appreciated.

With every good wish, I am

Sincerely,

Herman E. Talmadge

cc: Honorable Gerson H. Lush
Director, Washington Office
Brussels International Exposition
Department of State
Washington, D. C.

1958 MAR 27 AM 11 24

U.S. COMMISSIONER GENERAL
BRUSSELS WORLD EXHIBITION
1958

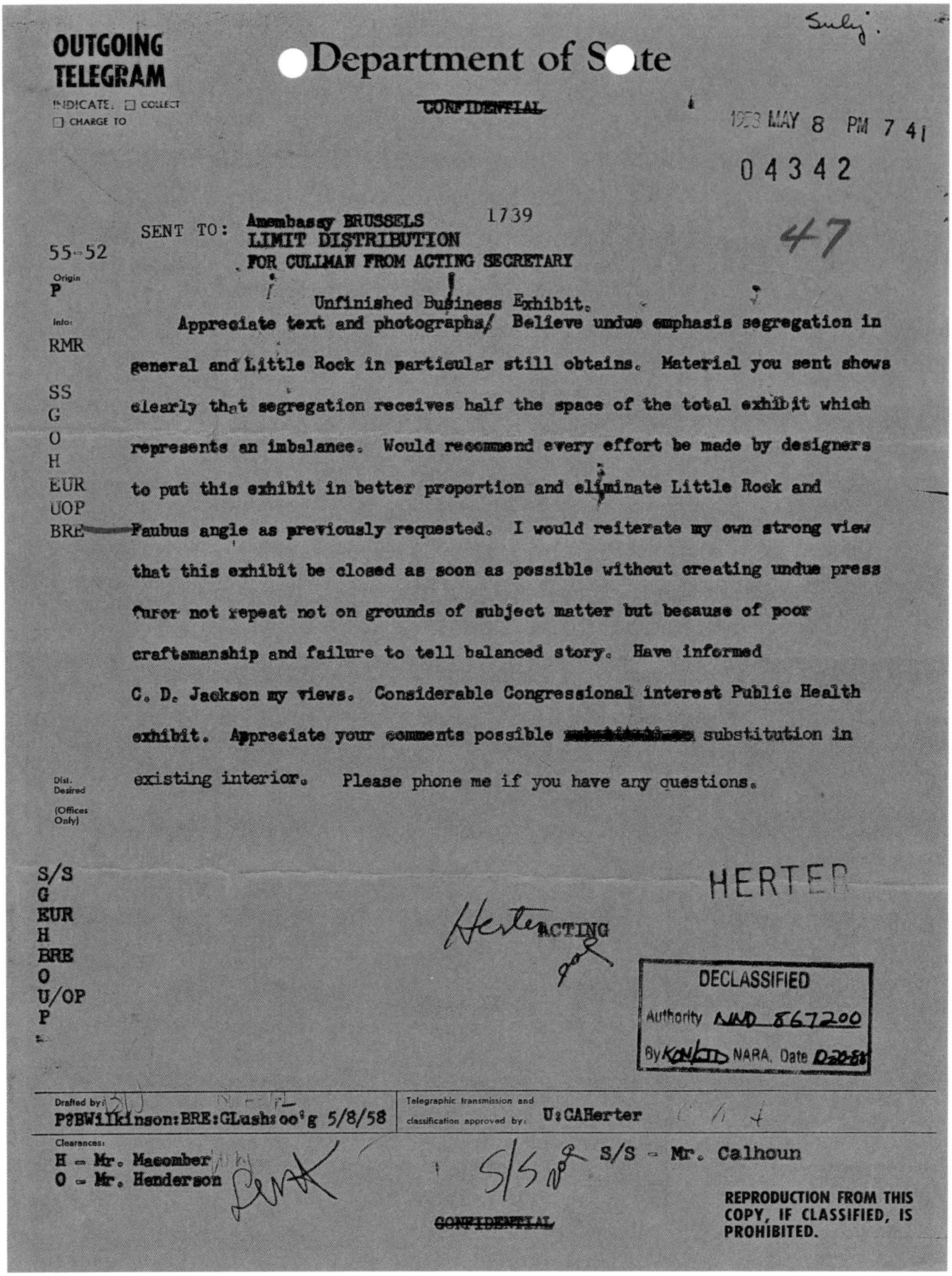
OUTGOING TELEGRAM

Department of State

INDICATE: ☐ COLLECT ☐ CHARGE TO

~~CONFIDENTIAL~~

1958 MAY 8 PM 7 41

04342

55-52

SENT TO: Amembassy BRUSSELS 1739
LIMIT DISTRIBUTION
FOR CULLMAN FROM ACTING SECRETARY

47

Origin P

Info: RMR SS G O H EUR UOP BRE

Unfinished Business Exhibit.

Appreciate text and photographs. Believe undue emphasis segregation in general and Little Rock in particular still obtains. Material you sent shows clearly that segregation receives half the space of the total exhibit which represents an imbalance. Would recommend every effort be made by designers to put this exhibit in better proportion and eliminate Little Rock and Faubus angle as previously requested. I would reiterate my own strong view that this exhibit be closed as soon as possible without creating undue press furor not repeat not on grounds of subject matter but because of poor craftsmanship and failure to tell balanced story. Have informed C. D. Jackson my views. Considerable Congressional interest Public Health exhibit. Appreciate your comments possible substitution in existing interior. Please phone me if you have any questions.

Dist. Desired (Offices Only)

S/S G EUR H BRE O U/OP P

HERTER

ACTING

DECLASSIFIED
Authority NND 867200
By KON/LTD NARA, Date 10-20-81

Drafted by: P:BWilkinson:BRE:GLush:oo'g 5/8/58

Telegraphic transmission and classification approved by: U:CAHerter

Clearances:
H - Mr. Macomber
O - Mr. Henderson

S/S - Mr. Calhoun

~~CONFIDENTIAL~~

REPRODUCTION FROM THIS COPY, IF CLASSIFIED, IS PROHIBITED.

Telegram from Acting Secretary of State Herter to US Embassy in Brussels ordering the exhibit's closing.

REPRODUCED AT THE NATIONAL ARCHIVES

DECLASSIFIED
Authority NND883136
By NARA Date 8/18/07

INCOMING TELEGRAM *Department of State*

55-51 CONFIDENTIAL

Action
P

Info
RMR
SS
G
O
H
EUR
UOP
BRE

Control: 20947
Rec'd: May 29, 1958
6:01 p.m.

FROM: Brussels

TO: Secretary of State

NO: 1687, May 29, 7 p.m.

FOR HERTER FROM PLAUT

DEPTEL 1761

Airmailing today photos Fortune exhibit which has been completely redesigned by Leo Leonni, Fortune art director. New exhibit opened to public for three days on limited basis to test reaction American and European visitors. Visits to exhibit presently controlled by trained guides who conduct visitors through in small groups. All reaction to date favorable. New exhibit not only eliminates all features heretofore considered objectionable but is well designed and well made. Must emphasize total change in atmosphere.

FOLGER

AB:GEV/4

DECLASSIFIED
Authority NND883136
By NARA, Date 6-2-93

CONFIDENTIAL

UNLESS "UNCLASSIFIED" REPRODUCTION FROM THIS COPY IS PROHIBITED.

US Pavilion notifies State Department the "Unfinished Business" has been modified and reopened.

ATOMIC AFTERMATH

The premature closing of "Unfinished Business" was not the only problem that overshadowed the end of the Brussels Fair. Independently of the official fair, an American entrepreneur, G. Robert Fleming, had brought to Belgium a Wild West Show, complete with cowboys and Sioux Indians, expecting to cash in on Europe's Hollywood-fed fascination with the myth of the Wild West. But the shows, immensely popular as they were, were badly planned: when a number of performances had to be cancelled because of rain, Fleming disappeared and the show collapsed into bankruptcy. While the cowboy star Gene Autry (who had hired out horses to the show) arranged to bring the cowboys and cowgirls back to the USA, the Indians were stranded. When, as Haddow explains, the Moscow Circus offered to put on a benefit performance for them, the US Embassy hurriedly stepped in to send them home, and to settle the debts incurred by the unofficial project. It was a public relations disaster, and one not without overtones of racism.

The US Government planned to spend between thirteen and fifteen million dollars on the Brussels exhibit. The most expensive show it had sponsored to date was the New Delhi Fair, which cost $622,000. The State Department proposed a whole chain of committees to supervise every aspect: there was even a committee of architects to supervise the architect. In advance of the Exposition they had sought opinion far and wide. But what all those consulted had in common, in most cases, was that they had never before considered how to represent America overseas.

Exhibits in the Soviet Pavilion at Brussels included industrial machinery, spacecraft models and heroic statuary.

EN 1913 LA PART
DE LA RUSSIE
DANS
LA PRODUCTION
MONDIALE
ETAIT EGALE
A 2,7 %
LA PART
DE L'U.R.S.S.
DANS
LA PRODUCTION
MONDIALE
REPRESENTE
ACTUELLEMENT
20 %
HET AANDEEL
VAN RUSLAND
IN DE WERELD
PRODUKTIE
BEDROEG IN 1913
2,7 PCT.
OP HET OGENBLIK
BEDRAAGT HET

The people who did have that experience were the seasoned professional designers, who were only brought in later, as with Peter Harnden. By planning from the top down, not the ground up, and by bowing to pressure for domestic reasons, the government came across as weak and evasive, and the strong aspects of the show, like the Children's Creative Center which was developed from the educational program of the Museum of Modern Art in New York, or the professional women's program, were pushed aside in the furor over "Unfinished Business." In the meantime, the monolithic Soviet exhibition contained the usual elements: statues of Lenin and the celebration of heavy industry, as well as a model apartment with a kitchen. But it was not to an apartment but a kitchen that attention was soon to turn.

The Soviet exhibit stressed the role of women in Soviet society.

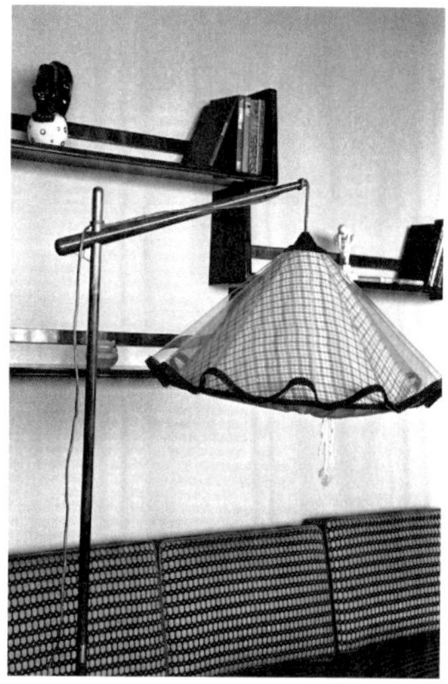

Russian dolls.
Soviet art.
Soviet lampshade.

A worker's "typical" kitchen and living room.

> Models of soviet aircraft.

—АЭРОФЛОТ—

HIGH NOON AT SOKOLNIKI PARK

AMERICAN NATIONAL EXHIBITION
MOSCOW, USSR
1959

In June 1957, the Soviet Party leader Nikita Khrushchev, with his customary bravura, issued an unexpected invitation to the USA for academic, cultural and scientific exchanges: he made the proposal during a live television interview on CBS's *Face the Nation.* While the method was obviously carefully calculated—such a public offer could hardly have been refused—what Khrushchev's full motives were are far from clear today and were just as puzzling at the time.

Before 1957, there had been a limited series of cultural exchanges involving groups of musicians and artists, as well as exchanges of a few scientific and technical exhibits. The Soviets had also invited Western nations to participate in a number of trade fairs, but the US never took part. The government was certainly open to cultural exchange, but was looking to stage a major American exhibition rather than participate in trade fairs. The State Department had plans for such an exhibition, which it presented to the Soviets in 1957, but the negotiations had gone nowhere. US policy on cultural changes had been set out in a policy document (NSC 5607) of June 1956, and had Eisenhower's strong support. According to Nicholas Cull, he urged the director of the USIA to offer the Soviets "unlimited access to America in exchange for same." This was what Khrushchev, seemingly spontaneously, was offering. If the first public official reaction was a blustering attack by Vice-President Richard Nixon on Soviet good faith, behind the scenes this invitation was recognized as an opportunity.

It is likely that Khrushchev felt the USSR would gain from industrial and technological contacts, just as it would gain from increased trade. The exchange of exhibitions and other cultural contacts would underpin his policy at a time of peaceful coexistence with the USA. As to what kind of exhibition he expected the US to mount, he had seen the Atomics exhibition in India in person and would have known about the other atomic energy exhibitions USIA had mounted, as well as the US Pavilions at other fairs. It has been suggested by Hixson that the inevitable display of American consumer goods that would result from a national US exhibition in the Soviet Union would force Soviet industry into improving its supply of consumer goods: there was certainly an attempt at improvement in the material culture in Russia subsequent to the exhibition. Caute suggests that the Moscow show stimulated demand for consumer goods and that later Soviet planning tried to provide for this.

< Vice President Nixon and Premier Khrushchev meet at the American Exhibition site.

Perhaps, as has often been suggested, Khrushchev was the last true believer: when he claimed, as he regularly did, that the supply of consumer goods in the USSR would overtake America's during the next decade, he did believe that a consumer society and a Communist society could coexist. Theoretically, a Communist society—one under the sway of a command economy—could indeed use its productive capacity to extend the range of consumer goods available rather than in a range of weapons for an armaments race. The capitalist model suggests that consumer choice is stimulated by demand which in turn is created by market forces and surpluses: the success of the American economy, and the rise in living standards in the postwar years, follows this model. But the American model is not the only one that can be used for comparison. Living standards rose elsewhere in Europe despite much closer governmental controls on credit and consumption, and even in comparatively closely regulated economies such as those in Scandinavia. So perhaps the notion of a Soviet Union rich in consumer goods is not as extreme an idea as it seems at first sight.

What kind of exhibition did the USA immediately envisage? American industry, severely retrenched in the Depression of the 1930s, had been reinvigorated through the production of war materiel. Spared Europe's challenge of rebuilding a devastated industrial sector after the end of the Second World War, the US economy had retooled itself for domestic production; by the 1950s, it was experiencing unprecedented growth. The products of industry and how they impacted on the lives of ordinary American citizens was a wholly valid subject. And, as will become clear, the question of what the US would exhibit was not decided through official government policy, but through the necessities and constraints both of budget and time.

In September 1958, an agreement was reached on a reciprocal exchange of national scientific, technological and cultural exhibitions. The Russians would exhibit in New York, the Americans in Moscow, both in the summer of 1959. This idea had Eisenhower's enthusiastic support. He looked forward to making a "spectacular" impact on Soviet society.

Eisenhower named George V. Allen, Director of the USIA, as the official coordinator of the American National Exhibition in Moscow; and appointed Harold "Chad" McClellan, former Assistant Secretary of Commerce for

International Affairs, to be general manager. McClellan had also been Chief Executive of the Old Colony Paint and Chemical Company of Los Angeles and a former president of the National Association of Manufacturers. He was a prominent Republican, and no stranger to the inner workings of Washington.

According to Eugene "Rocky" Staples, Deputy General Manager of the Exhibiton, the State Department would provide policy guidance, USIA would have the lead role in planning, and the Department of Commerce would have responsibility for the construction and management of operations; the project received an initial appropriation of $ 3,300,000 from the Mutual Security Program. This was a source of government funds that was controlled directly by the President. In fact, McClellan's memoir shows that he met with senior Congressmen—and in particular John Rooney, whose dislike of the USIA had led to earlier budget cuts—to gain support for the project. Once McClellan was confirmed as the American National Exhibition's General Manager, his first task was to negotiate a detailed agreement with the Russians.

The negotiations were long and complex, with the Soviet side finding issues and raising problems at every turn. McClellan, who led the American side of the negotiations, brought experience in both industry and government to his task, but he confirmed, when reporting to the President after the detailed accord was finalized, that it had been one of the toughest negotiations in which he had ever been involved. Both parties were aware that the deal was a reciprocal one, and that each side could create matching difficulties for the other. Even so, McClellan recalled that it was to be weeks before the Americans would realize that the tenacity of the Russian negotiators had put the American effort in serious jeopardy.

The Americans had proposed a fully equipped modern building, the New York City Coliseum, as the venue for the Soviet exhibition. Similar facilities did not exist in the Soviet Union. As a site in Moscow, the Russians first proposed Gorky Park in the center of the city, but the structural condition of the buildings to be used was doubtful. D. G. Borisenko, chairman of the Russian negotiating team and Deputy Chairman of the Soviet All Union Chamber of Commerce, then proposed another site, Sokolniki Park, some fifteen minutes by subway from the center of the city, and which had neither a suitable building in which to mount an exhibition nor paved roads.

McClellan notes that the Americans did their best not to show how pleased they were with his proposal. At the time the Exhibition Exchange Agreement was signed in September 1958, USIA had hoped that Buckminster Fuller would be able to design a large geodesic dome for the American Exhibition in Moscow, repeating the success of the Kabul dome. When, on October 23, 1958, the US team tentatively agreed to accept the Sokolniki Park site on the condition they could put up their own buildings, the Russians balked. Then, on November 6, 1958, the Russian team, perhaps fearing retaliatory delays by the Americans with the opening of their own reciprocal exhibition in New York, agreed not only to allow the construction of a dome and other buildings, but promised to buy the dome from the Americans at the close of the Exhibition.

It would not be the first time the Soviet people saw an exhibition in a park: as Caute has pointed out, the VDNKh, or Exhibition of the Achievements of the People's Economy, was established on a 600 acre park in Moscow, and was as important an attraction as Lenin's tomb, hosting eight million visitors every year. Each Soviet Republic had its own pavilion, leading to an excess of architectural styles and an effect described by Caute as "the motherland in miniature, the Disneyland of socialism in one country."

To solicit American private sector support for the Moscow effort, Eisenhower, a long-time advocate of cooperation between business and government, invited a group of business leaders to a lunch at the White House at which the plans for the exhibition were presented. Many in the group promised material and financial support of all kinds: items for display from automobiles to automatic blenders, from cutlery to *haute couture.* They not only promised the exhibition's contents, but also agreed to pay the cost of shipping them to Moscow. Such private sector support would avoid asking Congress for a large appropriation, which would otherwise have been essential and difficult to negotiate in the time constraints. While Congress had appropriated 14 million dollars for the State Department's US Pavilion at Expo '58, it had also cut the USIA's own funding very severely.

Eisenhower's enthusiasm for the project was matched by McClellan's, despite the endless negotiations with the Russians before an agreement was reached. Another eager supporter of the exhibition was Llewellyn Thompson, the US Ambassador in Moscow from 1957 to 1962. He believed that the exhibition could be an explosive success just by presenting the wealth of consumer goods available to ordinary people in the USA, without too many "propaganda objectives."

And a wealth of consumer goods was what the Soviets were going to see, even though it meant, as McClellan recalled, "creating a major exhibition facility in the heart of a forest in winter time on a frozen land; all of this in the heart of a primitive country 5,000 miles away from the U.S.A." Indeed, the delays in negotiating the site had put the budget under strain and the design process under intense pressure. How then was an exhibition and the buildings to house it to be planned, designed and installed in only seven months?

Entrance to the American Exhibition.

> Charles Eames (left), George Nelson (center) and Jack Masey with a model for the seven-screen film.

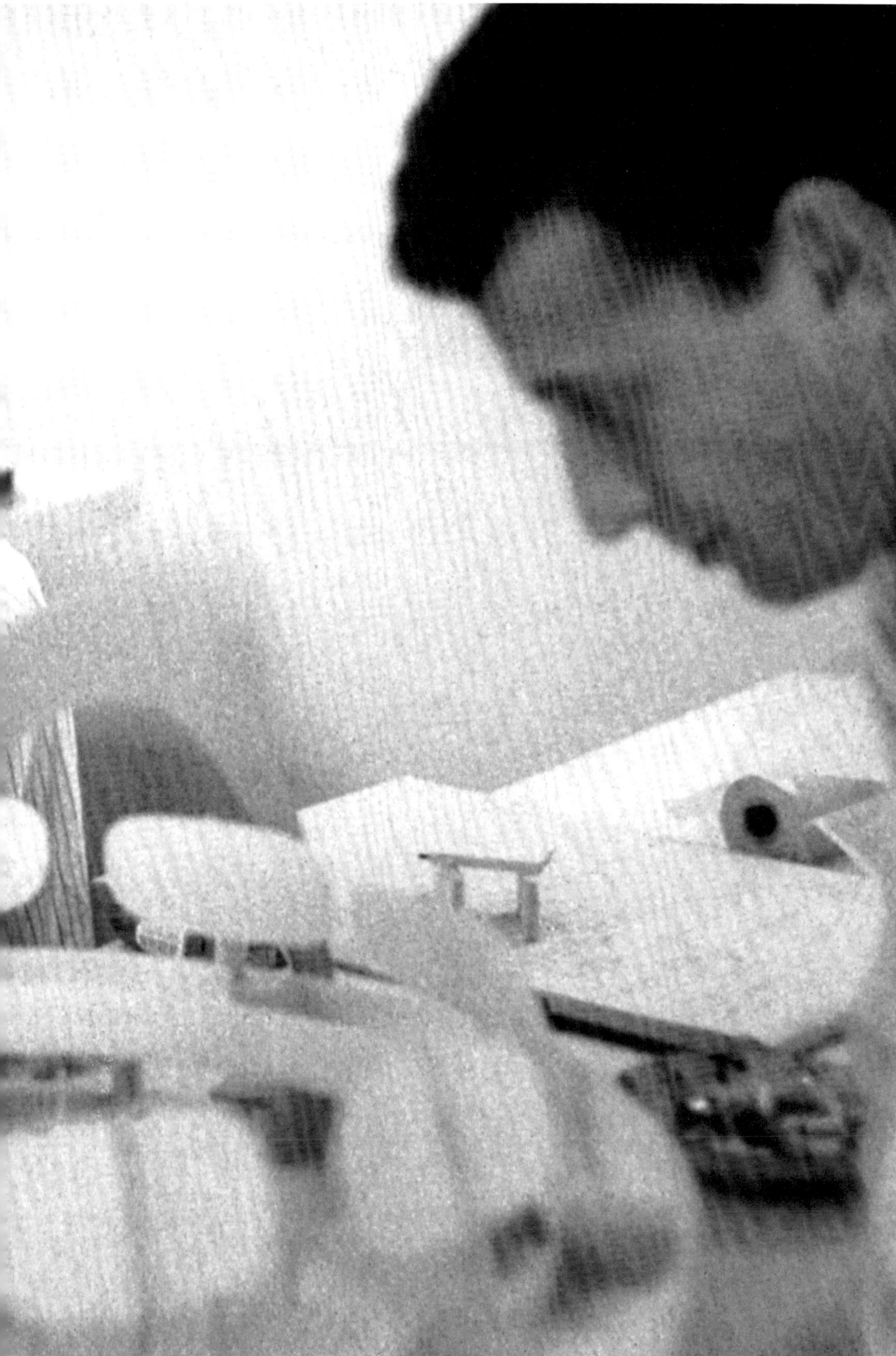

THE DESIGN PROCESS

George Nelson was born in Hartford, Connecticut in 1908. After studying architecture at Yale he won the Prix de Rome, and spent his time in Europe meeting and learning from some of the pioneers of Modernism in architecture and design. On his return to America he became associate editor and later consultant at *Architectural Forum,* where he introduced readers to many leading European architects and defended Modernism; during this time he eventually set up his own successful design practice. His project for a "storage wall" caught the eye of the owner of the Herman Miller Furniture Company, D. J. De Pree, who in 1945 invited Nelson to become design director. Nelson recommended that Herman Miller work with the designers Charles and Ray Eames, who for many Americans defined the postwar "California style" of American design. The Eameses' recliner fast became a design icon, and their work for Herman Miller would redefine the American office. Their own house in Pacific Palisades, California, has become a place of pilgrimage for designers, and their exhibition and film work influenced and continues to influence generations of American designers.

In September 1958 USIA named Jack Masey as Chief of Design and Construction of the American National Exhibition in Moscow. He immediately asked Buckminster Fuller to design a dome for Moscow: after the success of the Fuller dome in Kabul two years before, it made sense to invite him. Masey next called George Nelson, who in turn suggested that Charles and Ray Eames also join the team. At an early planning meeting between Nelson, the Eameses and Masey, it was decided that the dome would be the central statement, functioning as a kind of "information machine," and a second building—the Glass Pavilion—would house the consumer goods and other objects that Nelson derisively referred to as "the stuff." Nelson soon hired the architect Richard Barringer to design a prefabricated steel frame structure to display "the stuff" within the pavilion: Barringer's structure, the consumer goods bazaar, was soon nicknamed the Jungle Gym.

The Glass Pavilion was designed by Welton Becket Architects, a well-known West Coast company that had worked on Henry J. Kaiser's Hawaiian Village Project, and had already been involved in building embassies and overseas offices for the State Department. They were appointed at an early stage, on McClellan's recommendation. In a contentious early meeting between McClellan, representatives of Welton Becket's office, and George Nelson, McClellan told Nelson and the Welton Becket team that Welton Becket would be the "architects of record" (a role which today would be termed "job architect") for the whole project, including the dome. The antipathy between Nelson's team and

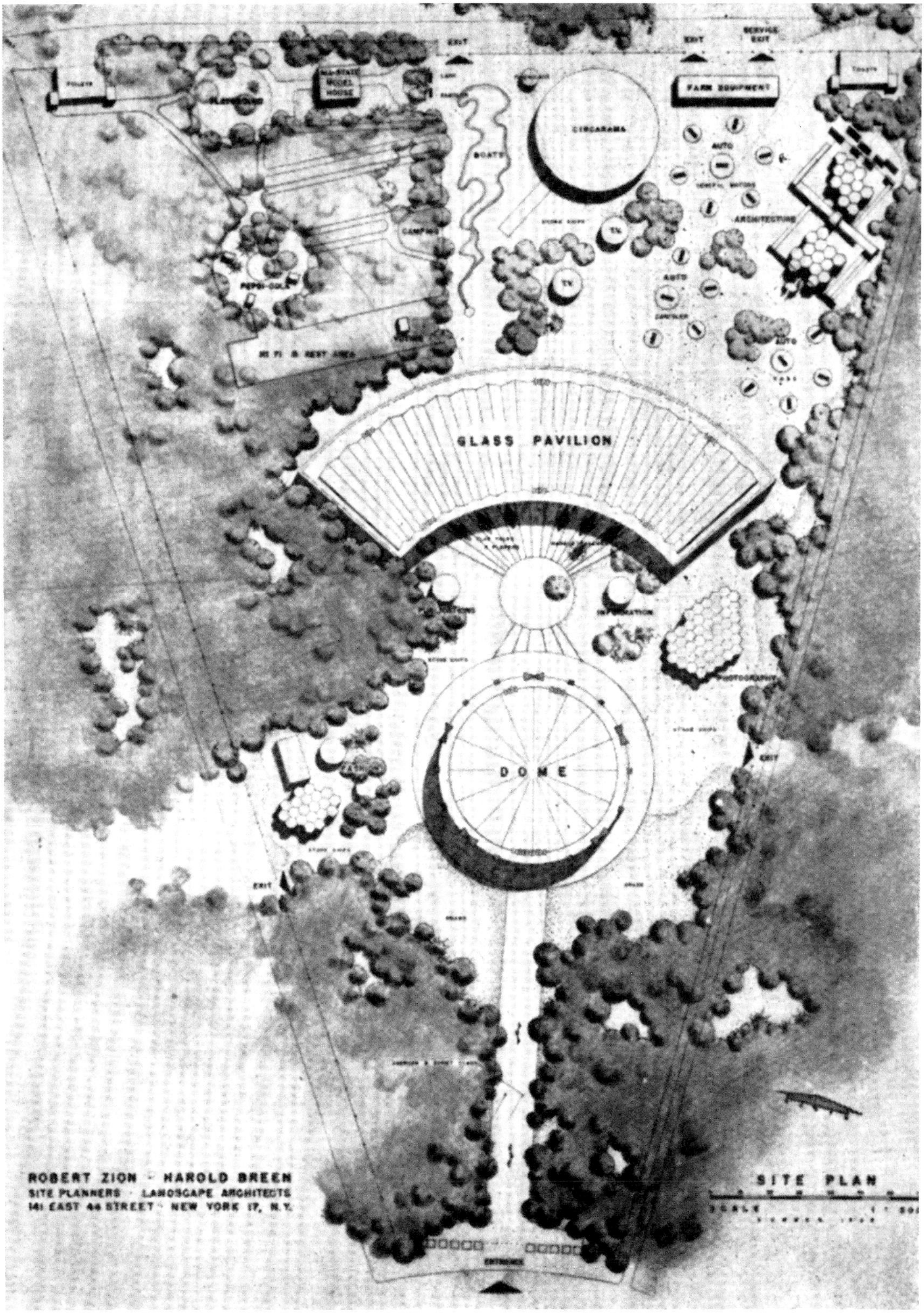

Plan of the exhibition site with the location of the dome and the glass pavilion.

Welton Becket soon became evident. Nelson's office designed the Jungle Gym, its intent being to provide the maximum flexibility for the display of (as yet undecided) objects and products, with optimum space for doing so. As a veritable "building within a building," the Jungle Gym had the effect of dominating the Glass Pavilion, the building that enclosed it.

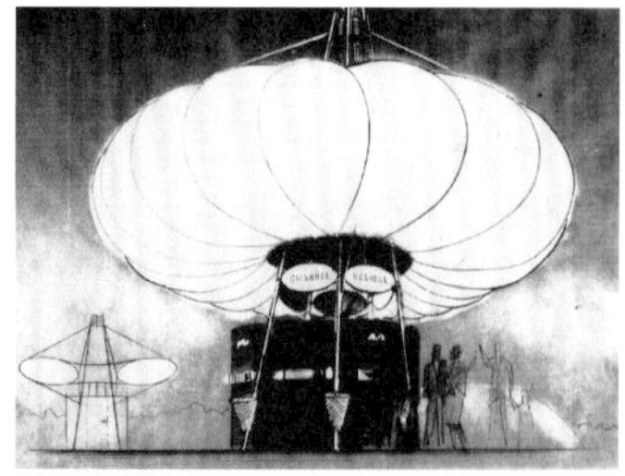

A third new structure was needed for the Exhibition, to shelter open-air presentations. Nelson designed this with Albert Dietz, a consultant engineer from MIT. As Philip George, Nelson's Project Manager, later recalled: "Nelson wanted something that was a tent or something like that, that was not fabric. So he started out with the idea of taking cubes of foam and carving it with a spoon. And he would carve these things out and then glue them all together and that would give you some kind of structure. Then we hired a man from MIT—Professor Dietz, a professor of engineering, and he came up with the idea of the acrylic umbrellas. It was a fabulous idea. But we had to prove they were structurally sound, so we assembled five umbrellas and borrowed [an] airplane and did a wind test, and they passed." These were named the Plastic Umbrellas.

Not only would the dome, the pavilion and the umbrellas need to be completed within seven months, but their contents also had to be selected, shipped to Moscow and installed in the same timeframe. Nelson's comments about this period were succinct: "... panics and problems. There were no funds. No budget. No program. No way of establishing a budget. No time for the design group to do its job (and no job for the design group to get started on.) No contract. No basis for a contract."

The Sokolniki Park exhibition site covered a seven-acre (2.83 hectares) triangular area, with the apex pointing towards the visitor entrance. In line with the main entrance was the welcoming building, the gold anodized dome, flanked by three open-air areas covered with clusters of inverted plastic umbrellas. One cluster covered a stage for a fashion show that included American dancing, a second covered "The Family of Man," an exhibition of photographs curated by Edward Steichen, and a third sheltered Peter Blake's large photographs of American architecture. Behind the dome was an information booth, and behind that the fan-shaped Glass Pavilion. Other exhibits—the current year's cars, voting machines, refreshment kiosks, and a "beauty kiosk"—were located around the main buildings. The model house, set up behind the Glass Pavilion, was split in half to allow the maximum number of visitors through it. In a tongue-in-cheek nod to the Sputnik, it quickly acquired the name "Splitnik."

Three proposals for kiosks on the Sokolniki Park site (unexecuted) by George Nelson.

REVOLUTION IN INFORMATION

It can be argued that there were two strands in postwar exhibition design. One strand was minimalist, springing from Bauhaus roots, and seen in the work of Herbert Bayer after his move to the USA, especially in his work for the Container Corporation of America. The second can be seen in the theories of London designer Misha Black, who had headed exhibition design for Britian's wartime Ministry of Information.

Black declared that "the essential function of a propaganda exhibition is to implant, or sustain, a general idea in the mind of the visitor." His theories of exhibition design, which advocated strong emphasis on the use of "actual physical objects" and demonstrations, contrasted with those of the Bauhaus typographer Herbert Bayer, whose approach could be termed architectonic: the organization of space through architecture and presentation of key ideas is what makes the impression on the visitor. Bayer, with László Moholy-Nagy and Walter Gropius, conceived the "field of vision" exhibition design technique that characterized the design of American and European exhibitions during the Cold War. George Nelson, writing in *Display,* described Bayer's "field of vision" approach: "If [images are] related to the height and angle of a person's vision, more can be seen simultaneously and with greater ease." (Significantly, in 1953, Nelson also noted the effectiveness of Bayer's theory in "wide-screen techniques such as Cinerama and Cinemascope.")

In 1952, at the invitation of Lamar Dodd, head of the Department of Fine Art at the University of Georgia in Athens, George Nelson created a proposal to bring together all elements of the art curriculum into a short but powerful presentation on communication. Nelson asked Charles Eames and Alexander Girard, a master of color and patterns, to collaborate on the project: they would combine Bayer's technique of "showing familiar material in unexpected ways" with Black's "actual physical objects" in a manner that had never been seen before. The presentation, made in 1953, was what the Eames office called "a rough sketch for a sample lesson for a hypothetical course" and George Nelson, more succinctly, Art X. Their intentions included "the breaking-down of barriers between fields of learning... making people a little more intuitive... increasing communication between people and things." This was achieved by a presentation that included graphics, film, slides, music, even synthetic smells, alongside a commentary by Nelson and Charles Eames. The idea was to build, through the onrush of different elements, an awareness of potential relationships between things that would otherwise be considered unrelated. Though it needed a crew of eight people just to operate the technology involved, as Nelson would comment, in the end "what counts is the force and clarity with which communication is established."

This approach was influenced by Claude Shannon's book *The Mathematical Theory of Communications,* which is generally regarded as the publication that launched communications theory onto a wider stage. In particular, Shannon took concepts such as noise and feedback used in audio and radio technology, and applied them to a range of other communication systems, which also allowed for an idea of positive noise. Some of the film elements presented in the "Sample Lesson" are based on the book, and these were developed into one of the Eameses' most enduring films, *A Communications Primer,* also made in 1953. John and Marilyn Neuhart, in their definitive book with Ray Eames on the Eames office, claim that "the lesson was a significant milestone in Charles' approach to presenting information in audiovisual form." If therefore, the "information machine in Moscow" required a multi-media approach, it had gone to designers more than ready to carry it out.

What is most interesting about the Sample Lesson is that it would seem also to be a reflection of the Eameses' own design practice. There is at first sight a contrast between the clear and simple forms of their furniture designs, which can be traced back to Charles's early collaboration with Saarinen, and the choreographed visual mixture of the many films and—later—exhibitions that they produced, with their rich outpouring of ingredients.

The multi-screen approach first tried out in the Sample Lesson was clearly derived from the editing technique of their films, which often use rostrum shots of still images as well as motion sequences. As a means of breaking down barriers and exploring connections in a rapid way that invited the viewers to find their own patterns of meaning, the Sample Lesson starts a process that achieved a new high in Moscow and was to go on to other, even larger projects.

Stanley Abercrombie, in his biography of George Nelson, noted that Nelson and Eames believed that carefully planned "bombardment" would communicate a large amount of information to viewers, but in the 1950s they were not the only designers who had begun to experiment with multiple dimensions and mixed media in exhibition design. Will Burtin had trained as a typographer in Germany, but left his homeland for the USA when he came under pressure to join Hitler's propaganda ministry as chief designer. Writing about his design practice, he comments: "In exhibitions, the adding of a third dimension plus physical motion allows full employment of the sensations of timing, scale, structure and

> *"The stakes are high, the budget low, the deadline impossible. We feel in this case how good our design is isn't important. Primarily, we want to make a simple sincere statement about American life."*
>
> George Nelson, *New York Herald Tribune,* March 11, 1959

volume." A believer in integration as a design principle, he was a master at presenting complex information, especially about the sciences, in two- and three-dimensional ways.

Burtin, who had designed the "Kalamazoo" exhibition for USIA in West Berlin in 1958, was one of the leading designers of exhibitions for the US government abroad, and had attended the annual cultural conferences initiated by Walter Paepcke in Aspen, Colorado. While Paepcke did indeed, as James Allen Sloan points out, move from an idealistic base in the late 1940s to a more business-oriented approach in the 1950s, the lists of panelists and speakers at the Aspen design conferences, compiled by Reyner Banham, were a virtual "Who's Who" of mid-century design.

Conceptually, the "information machine" in the Moscow Dome, and more specifically, the Eameses' seven-screen film, used the most cutting edge media techniques of the mid-20th century. Significantly, for designers working in today's new media, the Eameses' film presaged a transformation from design as being product-related to design as information-based. This shift of emphasis anticipated the wider economic shifts of what was later to be called the "information revolution," in which services began to replace manufacturing as the base economic model, and the potential of information systems to define and create markets began to be explored.

> Architect's rendering of the American Exhibition site showing the dome and the Glass Pavilion.

VAL.

THE DOME

There is one mystery concerning the Moscow Dome. Although Buckminster Fuller has for nearly half a century been given credit for its design, no evidence in the form of architectural drawings attesting to Fuller's role in the design of the dome appear to exist. Rather, a set of architectural drawings stored at the US National Archives are from Welton Becket and Associates.

The story of the Moscow Dome begins in late 1958 when Jack Masey was named Chief of Design for the upcoming exhibition in Moscow—some ten months before its scheduled opening. The first person he called was Fuller. When Masey asked Fuller if he could provide a dome for the Moscow exhibition, Fuller said, "Yes!" However, before the US Government could formalize an agreement with Fuller, site negotiations in Moscow bogged down. A final agreement between the US and the USSR was not signed until December 29, 1958: it confirmed that the Soviets would purchase the dome at the close of the exhibition.

Fuller had patented the concept of the geodesic dome in 1951 (Martin Pawley has pointed out that this was "the most important patent application of his life.") He had subsequently licensed a number of companies to exploit the patent, including the Pease Woodwork Company in Ohio, who produced a 12-meter plywood dome for use as a house from 1960 onwards. Fuller himself bought and lived in one for several years. More importantly, Kaiser Aluminum had also become a licensee. In 1946, Henry J. Kaiser had purchased three aluminum plants in the US; his visionary applications for aluminum allowed him to turn these former military production facilities into one of the largest aluminum-manufacturing companies in the US.

The Kaiser Aluminum and Chemical Company had provided the aluminum for Fuller's Ford Rotunda Dome in 1953. Recognizing the potential of Fuller's geodesic domes for his company, Kaiser licensed Fuller's geodesic patent and hired, with Fuller's blessing, Donald L. Richter, who had been taught by Fuller at the Institute of Design in Chicago, as well as at Black Mountain College (and who had been hired as Fuller's assistant when Fuller was teaching at MIT), to assist in their fabrication.

According to *Technology* magazine, "By 1956, Richter had convinced himself that an entirely different kind of geodesic dome could be built, combining both frame and skin into a single structural element. This was a marked departure from the basic Fuller dome, which consisted of two elements, a frame and a superimposed skin which were erected separately." Richter's system altered the relationship between a geodesic dome's frame and skin: with Fuller's system, the frame alone bore one hundred per cent of the load. With the system that Richter had patented, the skin bore seventy-five percent of the load and the frame bore only twenty-five percent. With Richter's patent, Kaiser could produce a new generation of domes with structural "hard-shells" that were capable of withstanding severe weather conditions.

When the Americans and Soviets belatedly agreed to build a geodesic dome in Moscow, Fuller knew that the dome type he had provided for the Kabul exhibition could not withstand a Moscow winter. Four years after the exhibition Fuller would recount in *Ideas and Integrities:* "They [the US] asked me which dome I would recommend, and I said a hard-shell dome as I wished to anticipate heavy snow loads in Russia … I recommended the Kaiser dome." The US Government took Fuller's advice. It commissioned Kaiser to produce its largest dome to date for the Moscow show. It also hired Welton Beckett and Associates to detail the dome's interior, entrances and periphery.

Fuller later came to Moscow to see the Kaiser dome, and wrote that "Khrushchev had said 'J. Buckingham Fuller must come to Russia to teach our engineers.'" When Buckminster Fuller describes himself as a designer, he is using the term in a slightly different sense to our current usage, for example in graphics or product design, where designers can go to great lengths to protect their copyrights. For Fuller, the term "design" implied more science than showmanship (though he himself was an effective showman). Seen in this sense, as Jane Pavitt has pointed out, his view of intellectual property was closer to that of an engineer than an artist. Fuller saw himself as "a comprehensive designer" whose function was "to anticipate what is going to be and to design with his eye on the timetable of an evolving industrial complex." Perhaps uniquely among 20th century designers, Fuller willingly and actively passed on what he knew for others to exploit.

At the time of the 1959 American Exhibition, however, the press credited Fuller with the design of its golden dome. Richter's contribution as its designer and Kaiser Aluminum's competence as its manufacturer were glossed over in press releases and news reports. In fact, the major portion of the credit for the design of the Moscow Dome belongs to Donald L. Richter, the inventor of its "Kaiser geodesics" hard shell.

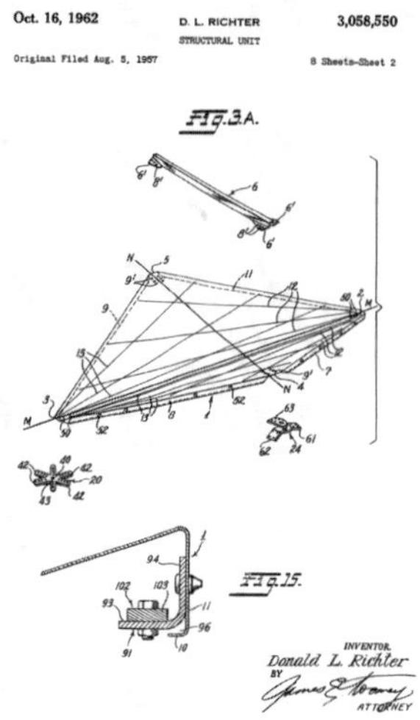

Richter's patent application for his "hard-shell" basic building unit.

> Plan, elevation and section for the Moscow Dome by Welton Becket Architects.

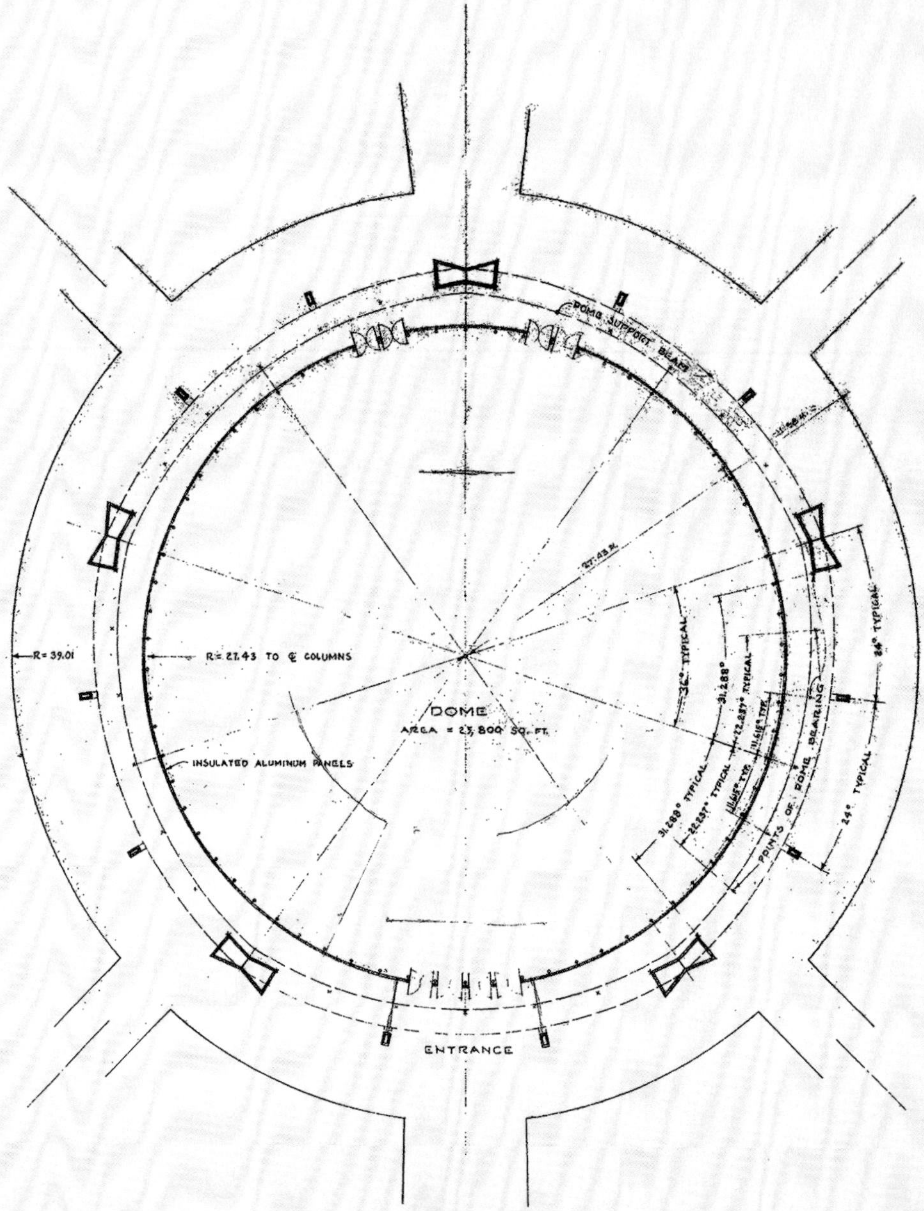

PLAN
SCALE 1:240

UNITED STATE

FOR THE AMERICAN NATIONAL EX

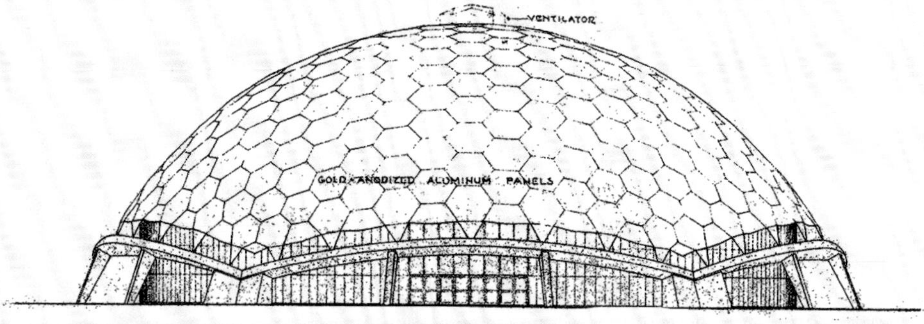

FRONT ELEVATION
SCALE 1:240

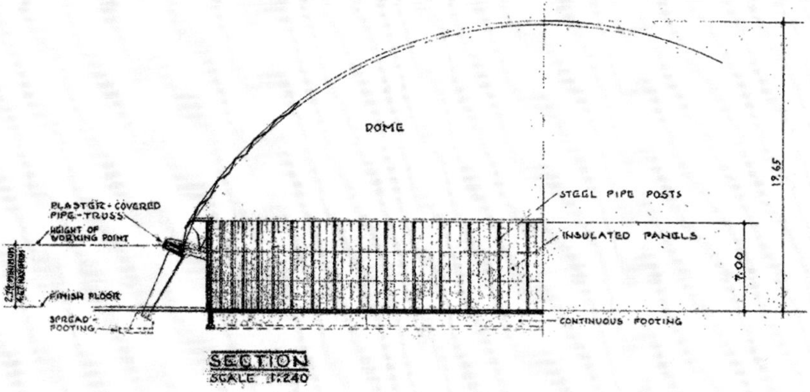

SECTION
SCALE 1:240

EXHIBIT
N IN MOSCOW, U.S.S.R.

WELTON BECKET AND ASSOCIATES
ARCHITECTS ENGINEERS
5657 WILSHIRE BLVD. LOS ANGELES CALIFORNIA

CONSTRUCTION

Peter Blake in his memoir *No Place like Utopia* talks eloquently of the problems of building in Moscow in midsummer. While the Russian construction crews were enthusiastic enough, the supplies of cement were wholly inadequate to form the floor slab for the Glass Pavilion. The available concrete, over-mixed with sand to compensate, disintegrated into choking dust. Whatever tools were on site frequently disappeared: Blake at one point had to go to Moscow's largest toyshop to buy up any hammers and trowels he could find. The heat was as oppressive as the dust; and the pace, especially in the last few weeks before opening, was hectic. When the guides arrived in Moscow in advance of the opening, they were drafted to work with George Nelson's team to help to install exhibits in the Jungle Gym: as one guide, George Feifer, said, "I could not imagine how this would open in four days—it was utter chaos!"

The Soviets regularly contrived delays—the fashion show was almost cancelled when their charter flight was refused permission to land in Moscow, not to mention the arrival of twenty-six additional paintings for the art exhibit (for which the space had already been filled). The Americans caused their own problems too. There was an on site restaurant open during construction for the staff. After a lunch that included George Nelson and some of his team, the Russian staff found some items of crockery had been embellished with drawings. Harold McClellan's unpublished memoir relates that a Russian official described one of them as "a naked figure of a woman: she was misshapen and ugly. She was holding the hammer and sickle … and the figure of the woman was the Soviet flag." The matter was so serious that the offending crockery had been passed directly to the Foreign Ministry.

Components arriving on site.

Raising the center of the Dome.

Working on the exterior of the Dome.

McClellan, suspecting at once who had made the offending drawing, was forced on behalf of the US to make a full apology (Nelson refused to apologize for the incident, but his habit of making quick comic drawings of naked women was well known.)

There were also construction problems: the Italian firm that supplied the structural steelwork for the Glass Pavilion confused tension and compression members, and almost all of them had to be replaced, as Philip George, Nelson's project manager on site, recalls. Because of the concrete shortage, the individual footings for the structure had to be poured before the entire floor was formed, instead of the structure resting on the concrete slab, as designed. "We were fighting schedules. It was frightening," George remembered.

McClellan's memoir details the difficulties and crises he encountered during the negotiations and the installation of the exhibition; his ability to manage a large team in what was often almost a hostile environment was admirable. That the Moscow Exhibition happened at all is a tribute to everyone involved. Compared to the Brussels Expo the year before, Moscow had one quarter of the budget and one half of the time to create two giant pavilions in an impenetrable country critically short of resources.

Completing construction work on the roof of the Dome.

The completed Dome.

> Workers applying final touches to the Kaiser dome.

"The special coating that protected the dome's panels during shipping had stuck on the panels. It looked awful! We had a plane-load of razor blades shipped to Moscow and we scraped the entire surface of the dome by hand! We never told anybody because it was so embarrassing."

Philip George, George Nelson's Project Manger

Two alternative schemes
considered for the film projection,
on four and nine screens.

THE DOME INTERIOR AND THE FILM

The "information machine," as Nelson termed the Dome, had three main components: the Eameses' film, the IBM RAMAC computer that could answer 4000 questions about the USA with pre-programmed responses in Russian, and a series of graphic information panels on agriculture, education, scientific research, and so on. The Eameses' seven-screen film, *Glimpses of the USA,* was the main feature. Its cinematic intention was to carry the viewer along on a tide of images through the journey of an American day. This was to be done through multiple inputs, not a single storyline. There was an editorial reason for this: America was a vast, diffuse and varied country. Going to school in New York was very different from going to school in Ohio or California. Going to work as a farmer, a factory worker or an office executive was different too. A single narrative would obscure the hundreds of other typical American days. The Eameses believed, from their experience with short films, that multiple, changing images conveyed as much, if not more information than single static ones. They were themselves inveterate collectors of images, and fascinated by multiple visual events such as parades and circuses (the subjects of some of their films). Projects such as "The Toy" and the "House of Cards" had already shown their versatility with wide ranges of images. The immense interior surface of the Dome was the perfect setting for a grand experiment in the potential of multi-screen as well as multi-image: their most ambitious similar project to date had been the three-screen slide show *Konditorei,* in 1955.

The plethora of images the Eameses were proposing would show to Russian visitors the deluge of consumer goods in the Jungle Gym in authentic American settings. Nelson may have complained ruefully about being landed with "the stuff" while the Eameses got to do the art, but in fact the overarching script was the same for both: create a visual proof of the abundance of American society. There is no exact match between the contents of the films and the rest of the display: such a twinning would have been parodic, but each closely reinforced the message of the other.

This was not something the Eameses had often tried before, and certainly not on such a scale (each screen was 20 by 30 feet, or 6 by 9 meters). There were many discussions between the Eameses, Nelson and Masey about the right number: design sketches show arrangements for four and nine screens. Seven was chosen as being the right number for avoiding "sensory overload," as Beatriz Colomina has pointed out. More than seven and the eye simply cannot keep up: fewer than seven and the depth of the viewer's cognition is too low: the viewer remembers the number of screens. If this sounds confusing now, it was more so at the time.

> The final seven-screen configuration during projection of the films.

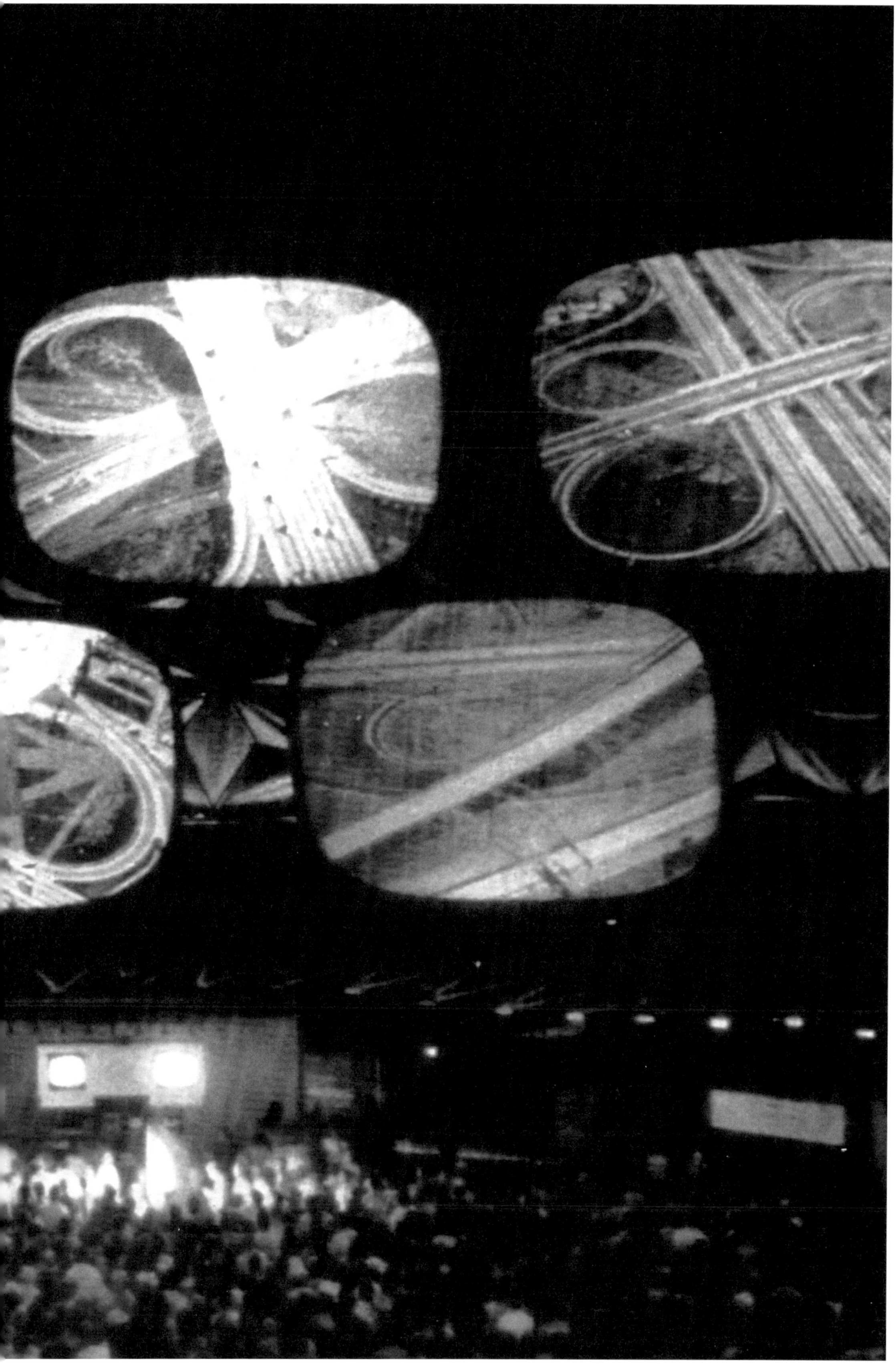

McClellan reports in his memoir that Charles Eames was invited to submit a screenplay or at least a synopsis in advance, which he resolutely refused to do, and continued to argue eloquently against defining his project during a two hour meeting. At the end, McClellan signed the contract: he was convinced, despite his business instincts, that Eames was the only person with the imagination to carry out the project, and the general response to the Eameses' collages of superhighways and school buses, milk bottles on doorsteps and barbecues on patios has been, since Moscow, one of respect for the groundbreaking way in which they handled and conveyed visual information. Jeffrey Meikle retrospectively identifies *Glimpses of the USA* as the pivotal moment when American design changed from being object-oriented to information-oriented.

Masey's eyewitness account describes the excitement in the Dome when the Eameses arrived in Moscow shortly before the opening of the Exhibition, hand-carrying seven cans of 35mm film: "The Radio City Music Hall's top projectionist was on loan to the Exhibition. He lost no time loading the projectors with the film the Eameses had brought. And so, that evening we—and Charles and Ray Eames, as well—were able to have our first full-scale testing of the multi-screen presentation of *Glimpses of the USA*. Word spread and a crowd began gathering in the dome to watch the preview. There were Russian construction workers, Finnish carpenters and American designers, all eager to see what magic would appear on the seven screens.

"Soon the projectors were synchronized and the twelve minute multi-screen film got under way. The combination of moving and still images, accompanied by a powerful musical score by Elmer Bernstein, was fantastic! Here were early morning scenes of children eating breakfast across America; aerial views of deserts, mountains and plains; cloverleaf highways jammed with automobiles; skyscrapers glistening at night; and the final closing scene: a bunch of Forget-me-nots—the universal symbol of friendship—depicted as a single image on the central screen. The Russians whispered "nezabutki" (Forget-me-nots in Russian) and there were tears all around.

"When the film showing ended there was silence. Everyone was mesmerized by the show. None of us had ever seen anything quite like it. Then the audience burst into wild applause."

Charles and Ray Eames in Moscow.

Design drawing for the screen system.

Films in projection.

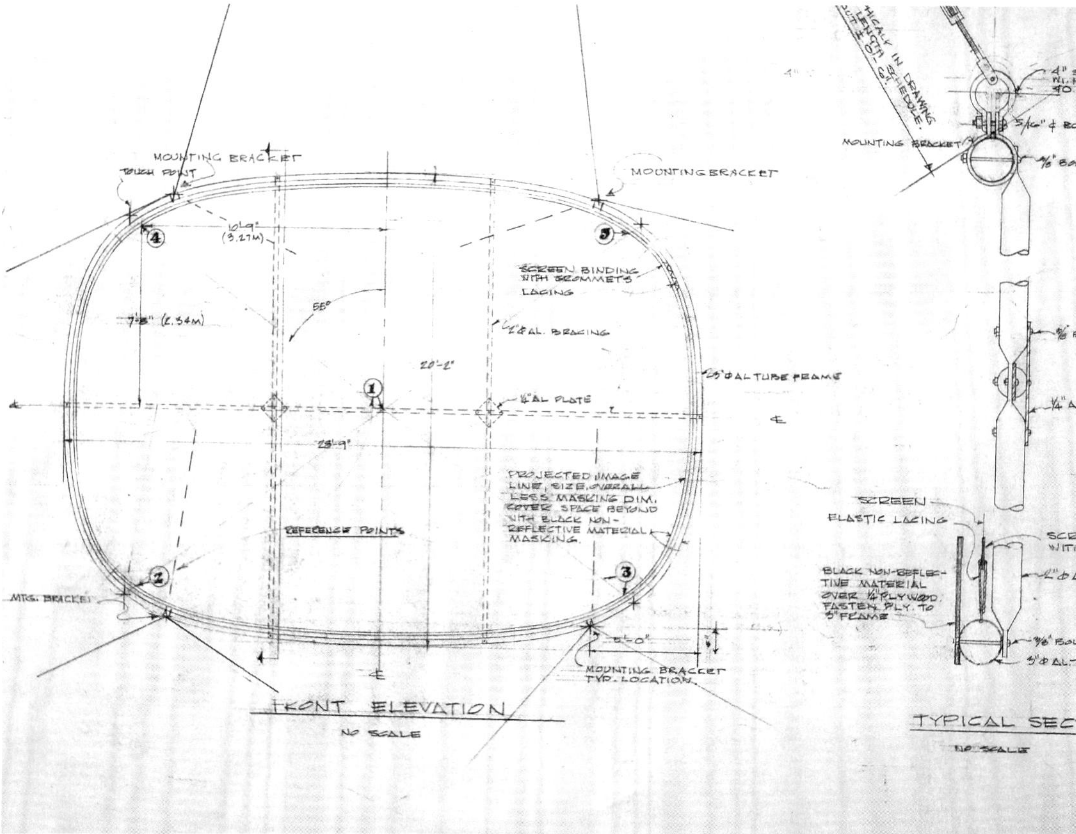
MOUNTING BRACKET
TOUCH POINT
MOUNTING BRACKET
(3.27M)
SCREEN BINDING WITH GROMMETS
LACING
7'-8" (2.34M)
55°
AL. BRACING
20'-2"
AL TUBE FRAME
AL PLATE
23'-9"
PROJECTED IMAGE LINE, SIZE OVERALL LESS MASKING DIM. COVER SPACE BEYOND WITH BLACK NON-REFLECTIVE MATERIAL MASKING.
REFERENCE POINTS
MTG. BRACKET
MOUNTING BRACKET TYP. LOCATION.
FRONT ELEVATION
NO SCALE
MOUNTING BRACKET
SCREEN
ELASTIC LACING
BLACK NON-REFLECTIVE MATERIAL OVER 1/4" PLYWOOD FASTEN PLY. TO 5" FRAME
TYPICAL SECT
NO SCALE

DESIGNING THE JUNGLE GYM

Richard Barringer's ingenious structure inside the Glass Pavillion not only doubled the available floor space by adding multiple levels for displays, it also provided a means of allocating exhibits within the whole. Barringer built a scale model of the entire structure in Nelson's office, which was used as a template for placing exhibits. It was quite literally a template: the model was shipped to Moscow in sections, and used to show the construction crews how the structure should be built, and, later, where the exhibits were to go. This method was employed because there was simply no time to prepare a complete set of drawings for use on site. This novel but necessary approach shows the resourcefuness of Nelson and his team under pressure.

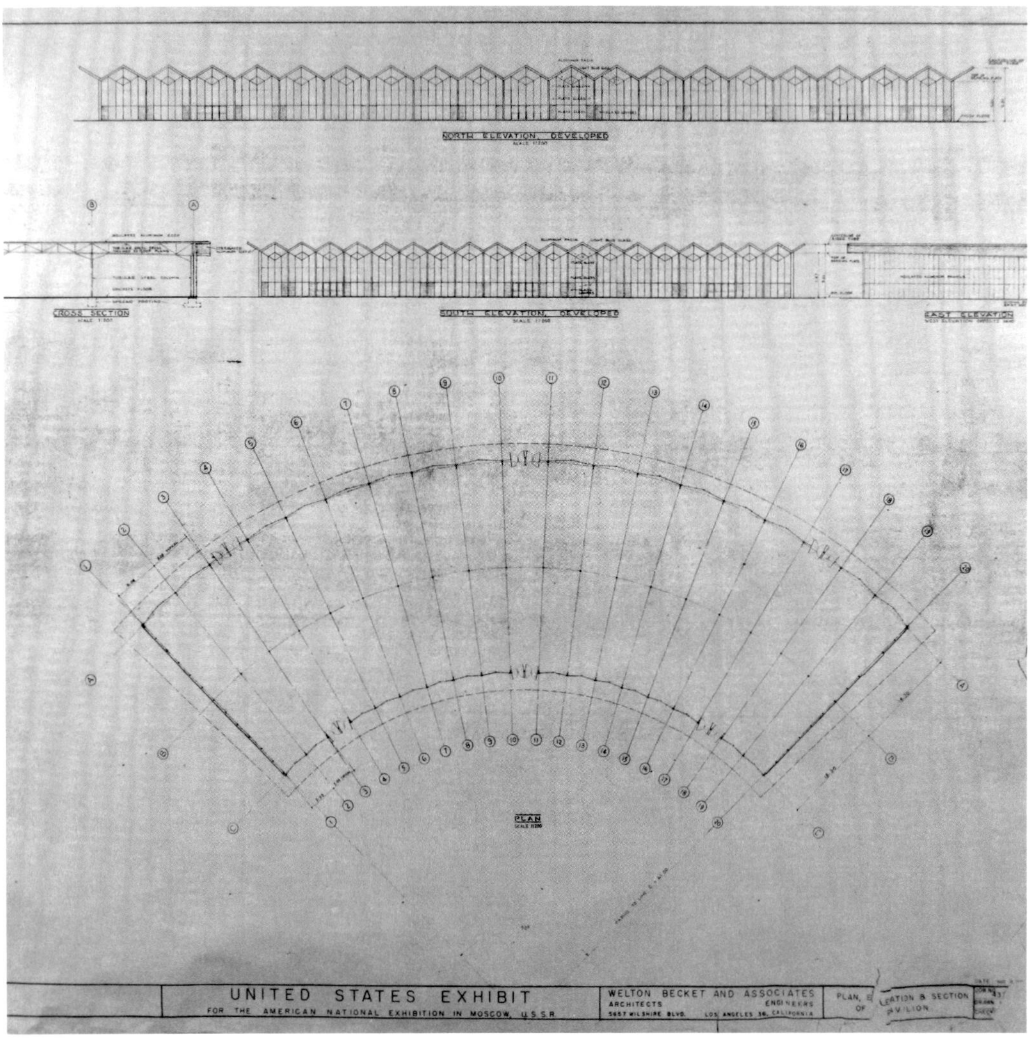

Plan, elevation and section for the Glass Pavilion building.

Two views of the scale model of the Jungle Gym built within the Glass Pavilion.

> Working on a scale model in the Nelson office.

TO OPEN
REMOVE THIS
SIDE ONLY

FILLING THE JUNGLE GYM

The intent of the USIA planners and Nelson's team was to fill the Glass Pavilion with "the fruits of American labors." There was a general understanding that the selection of items for display should be as multitudinous and varied as possible, but USIA were dependent on the largesse of the private sector in acquiring these items.

The floor area of the Jungle Gym was, according to Philip George, "the size of a city block." To fill it would be the equivalent of filling two floors of a major department store, such as Selfridges in London or Saks in New York. Barringer's design allowed individual spaces to be allotted to different product areas, and for some of these, individual companies or industry committees agreed to take responsibility for the design, supply and installation of exhibits.

George Nelson during the installation of the Jungle Gym.

Plan of the Jungle Gym structure.

Installing exhibits in the Jungle Gym.

However, for most of the contents of the Jungle Gym, Nelson and his team had to source suitable material, and persuade the manufacturers not only to donate the products but to ship them to Moscow as well, though the procurement staff of the OITF also helped with this part of the process.

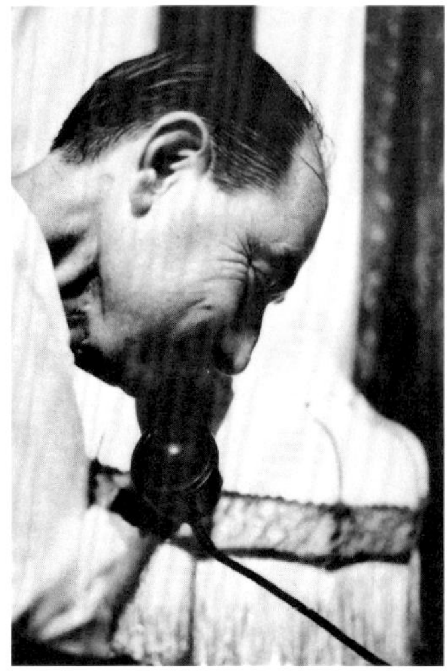

George Nelson's design office was in the same building as *Industrial Design* magazine, which each year published a review of the best furniture and product design of the year. Faced with the task of filling the Jungle Gym, Nelson's team turned to *Industrial Design*'s editor, Jane Fiske Mitarachi, who provided them with the magazine's lists of potential suppliers. She also suggested that one of their writers, Irma Weinig, join Nelson's team to help. Nancy Friday, one of Nelson's assistants, was the other member of the team. Their task was an enormous one, as Philip George recalled: "The two ladies really researched and found the sources that we wanted. Then we would select. If she got five different sources of different designs, we would then sit down and say 'That's the one I want, and if you can't get that, then this is number two.' We had to rely on them." In addition to the exhibits of consumer items, Lucia DeRespinis, the sole female industrial designer in Nelson's office, designed and furnished the model apartment in the Jungle Gym (it was intended to represent a home for an American physician with one child). Others in the office were also responsible for laying out the art exhibit and the exhibition of books and newspapers.

For those displays provided by individual companies or committees, the Nelson office would allot space and provide text and graphics, while the specific design was undertaken by the company concerned. These included RCA for the color television studio, Singer for industrial and domestic sewing machines, Whirlpool for the kitchen, and Grand Union for the supermarket. The fashion and fabrics display in the Jungle Gym was funded by a committee from Bergdorf Goodman and designed by Tom Lee Ltd; the same committee organised the outdoor fashion show. Working with these disparate companies meant it was not possible for the design team to apply a uniform design policy.

Nelson's efforts and the support of manufacturers did fill the Jungle Gym with a cornucopia of "stuff": toys, sports equipment, domestic appliances, hi-fi equipment, TV sets, cameras, clothing and textiles ("Levi Strauss gave us stacks of Levis," Philip George recalls, "and they were stacked up in piles. And they all disappeared. We wanted them to be stolen.") There was no time to first bring all this material to New York and to make a definitive selection and space allocation: the exhibit items went straight to Moscow. Sometimes, as Irma Weinig points out in an *Industrial Design* article, the manufacturers would choose to send a later

< Installing exhibits in the Jungle Gym.

George Nelson.

or a different model from the one promised, and "sometimes they would let us know that, sometimes not."

Twelve members of Nelson's office had arrived in Moscow by June 23rd, a month before the exhibition was to open. "The consensus was that we were six weeks behind schedule," Irma Weinig wrote. "The Nelson staff designed many of the exhibits on the spot, and did the actual installation—hanging objects, texts and panels with the sometime help of the guides and some Marines from the US Embassy. Charles and Ray Eames pitched in. This was practically a hand-made operation." According to Philip George, "we only knew two or three weeks before the opening what was definitely going to go into the Jungle Gym."

That the exhibition opened at all is astonishing. It had been assembled from scratch—both buildings and contents—in under seven months, had virtually no budget for acquiring exhibits, and was subject to major construction delays. Many of the final choices were left to George Nelson and his team, or to the manufacturers themselves who contributed material. One cannot argue from such a process of happenstance that there was an overriding policy of showing "modern design," save to the extent that the taste of those making the selection, young designers working in New York, would have been for the modern and the contemporary rather than the traditional (although historic items, for example from Colonial Williamsburg, were included in the displays).

According to Irma Weinig, "The products were chosen to demonstrate quality, finish and ingenuity, the range in size or different needs, and the abundance which gives the consumer such wide choice … To Russians who parroted Pravda's cry 'Where are the machines and techniques for whch the US is famous?' the answer was 'it is taken for granted that those machines and techniques are behind these products.' The entire exhibit was directed to the average Russian, not the educated or technically trained, and aimed to put him squarely within the American frame."

Model shot of the Plastic Umbrellas used for outside exhibits.

The finished Plastic Umbrellas near the Dome.

Installing and erecting the Plastic Umbrellas.

> Visitors arriving on the opening day of the Exhibition.

THE GREAT DEBATE

The intense interest that the Moscow and New York exhibitions had created in the American press, fuelled in part by plain curiosity about the Soviets, and in part out of deep concerns about the Space Race, meant that the opening event in Moscow would be not just a formal exchange of diplomatic greetings. Nixon was eager to attend the opening, and Eisenhower agreed, sending with his Vice-President a personal letter to Khrushchev inviting him to visit the USA. Nixon saw the opportunity in a personal meeting with the Russian leader to boost his own anti-Communist stance back home; the Russian view, as Fursenko and Natali explain, was that Nixon had replaced Dulles as Eisenhower's most hawkish advisor, seeking to thwart détente. Nixon found Khrushchev "a man of great warmth and totally belligerent"—an assessment that was perhaps reflected in the mood swings of the debate that took place between them.

While the major confrontations between Khrushchev and Nixon took place primarily at two sites—a 16-minute exchange in the television studio in the Glass Pavilion, and a forty-five minute exchange in the model house kitchen—the debate between the two is generally referred to as the "kitchen debate." On their walk through the US site, they also bypassed the voting machines (which Khrushchev did not like) and stopped at a kiosk dispensing Pepsi-Cola (which he did).

Many have speculated that Nixon went to Moscow intending to confront Khrushchev: Nixon always took a hard line with the Communists, and the Republicans were doing badly in the polls, so he had everything to gain. The confrontation between the two leaders was first evident in the television studio, where a recording of it was made on an early Ampex videotape recorder. According to McClellan, "Premier Khrushchev spoke first in the television studio ... He attempted to persuade his listeners that Russia was, in most departments, ahead of the United States." This sixteen-minute tape was flown back to the USA, and was seen on American television in late July. The original tape has been lost: the copyright version in the Library of Congress is a studio copy, as Jim Lindner has pointed out. The loss of a document of such importance may seem astonishing, but it was, as Lindner points out, a risk endemic in the way early videotape was used and reused to save costs. There is also an amateur 8mm film record of the meeting in the television studio, probably made by one of the technical team. It has no soundtrack.

Nixon in his memoirs suggested that he had chosen the kitchen for the debate (if a discussion that veered between comment and insults can be called a debate), because it was one of the most popular exhibits in the exhibition. Though how he can have known on opening day what would be popular he doesn't explain.

The image of Nixon and Khrushchev together in the kitchen of the Splitnik has become a Cold War icon. A kitchen setting may seem oddly domestic, but the Soviets had exhibited a "typical worker's kitchen" at Expo '58 in Brussels, though few Soviet workers might have been able to afford such luxury. Certainly some of the remarks exchanged that became well known were not necessarily spoken in the kitchen. For example, Khrushchev's comparison of standards of living and material progress—"When we catch you up, in passing you by, we will wave to you"—seems to have been made in the television studio. The conversation as reported in the *New York Times* seems more like a series of impromptu exchanges, not following a single thread.

The Kitchen Debate, for all the trivial asides and non-sequiturs of some of the remarks, remains one of the central events of the Cold War. That it received such wide attention at the time reminds us that both superpowers claimed the cultural sector as their own domain, and that the disagreements were not just about ideology and military power, but concerned material prosperity and consumer technology as well. The idea of the President of the Soviet Union and the Vice-President of the USA arguing over a washing machine, then agreeing to dislike jazz, does have a certain surreal quailty to it, but it also demonstrates how both, however much they might distrust one another, were looking for a way of avoiding a military confrontation. It was in some ways a unique encounter: in an informal setting, unscripted, and personal, taking place live and in public. The Kitchen Debate marked the start of something new.

Nixon and Khrushchev
in the television studio.

Premier and Vice-President
in the Jungle Gym.

The debate continues in the television studio and outdoors.

THE KITCHEN DEBATE

...

Khrushchev: **We want to live in peace and friendship with Americans because we are the two most powerful countries and if we live in friendship then other countries will also live in friendship. But if there is a country that is too war-minded we could pull its ears a little and say: Don't you dare; fighting is not allowed now; this is a period of atomic armament; some foolish one could start a war and then even a wise one couldn't finish the war. Therefore, we are governed by this idea in our policy—internal and foreign. How long has America existed? Three hundred years?**

Nixon: **One hundred and fifty years.**

Khrushchev: **"One hundred and fifty years?" Well then we will say America has been in existence for 150 years and this is the level she has reached. We have existed not quite forty-two years and in another seven years we will be on the same level as America. When we catch you up, in passing you by, we will wave to you. Then if you wish we can stop and say: Please follow up. Plainly speaking, if you want capitalism you can live that way. That is your own affair and doesn't concern us.**

Nixon [pointing to American workmen]: **With men like that we are strong. But these men, Soviet and American, work together well for peace, even as they have worked together in building this exhibition. This is the way it should be. I can only say that if this competition in which you plan to outstrip us is to do the best for both of our peoples and for peoples everywhere, there must be a free exchange of ideas. After all, you don't know everything.**

Khrushchev: **If I don't know everything you don't know anything about communism except fear of it.**

Nixon: **There are some instances where you may be ahead of us, for example in the development of the thrust of your rockets for the investigation of outer space; there may be some instances in which we are ahead of you—in color television, for instance.**

Khrushchev: **No, we are up with you on this, too. We have bested you in one technique and also in the other.**

Nixon: **You see, you never concede anything.**

Khrushchev: **I do not give up.**

Nixon: **You must not be afraid of ideas.**

Khrushchev: **We are telling you not to be afraid of ideas. We have no reason to be afraid. We have already broken free from such a situation. You are a lawyer for capitalism and I am a lawyer for communism. Let's compare.**

Nixon: **The way you dominate the conversation you would make a good lawyer yourself.** [Halting Khrushchev at model kitchen in model house]**: You had a very nice house in your exhibition in New York. My wife and I saw and enjoyed it very much. I want to show you this kitchen.**

Khrushchev: **We have such things.**

Nixon: **This is the newest model. This is the kind which is built in thousands of units for direct installation in the houses. … Let me give you an example you can appreciate. Our steelworkers, as you know, are on strike. But any steelworker could buy this house.**

Khrushchev: **We have steelworkers and we have peasants who also can afford to spend $ 14,000 for a house. We build firmly. We build for our children and grandchildren.**

Nixon [pointing to television screen]**: We can see here what is happening in other parts of the home.**

Khrushchev: **This is probably always out of order.**

Nixon: **Da (yes).**

Khrushchev: **Don't you have a machine that puts food into the mouth and pushes it down? Many things you've shown us are interesting but they are not needed in life. They have no useful purpose. They are merely gadgets. We have a saying, If you have bedbugs you have to catch one and pour boiling water into the ear.**

Nixon: **We have another saying. This is that the way to kill a fly is to make it drink whisky. But we have a better use for whisky.** [Aside]
I like to have this battle of wits with the Chairman. He knows his business.

Nixon [hearing jazz music]: **I don't like jazz music.**

Khrushchev: **I don't like it either.**

Nixon: **But my girls like it.**

Khrushchev [apologizing]: **I always speak frankly. (He said he hoped he had not offended Mr. Nixon.)**

Nixon: **I've been insulted by experts. Everything we say is in good humor.**
...

A tense moment between the two leaders in the Glass Pavilion.

> The crucial confrontation: Nixon and Khrushchev in the Splitnik kitchen.

>> The Glass Pavilion awaiting the opening day.

PICK·A·P

SOS

WELCOME TO THE USA

What was offered by the USA to the people of the Soviet Union was more than a cornucopia of consumer goods: the American Exhibition was an attempt to create a kind of "tourism in reverse" experience for Russian visitors, most of whom would never be able to visit the United States. As former USIA officer Yale Richmond notes, Norman Cousins has described the American Exhibition at Sokolniki Park as one that focused on "the individual citizen—how he goes shopping, how he has his hair cut, what the inside of his home is like, what he does for work and what he does for fun. In short, what everyday life is like for many millions of people who live in the U.S."

In addition to the displays of consumer items in the Jungle Gym that were linked in a certain fashion to daily life in America by the Eames film in the Dome, the Glass Pavilion also housed the highly controversial art show. Exhibits such as the RAMAC electronic brain, the television studio and the Whirlpool Miracle kitchen demonstration area were intended to provide Russian visitors with interactive or hands-on experiences. Russian visitors were free to touch shoes and clothing, to read books, to pick up objects, and to open refrigerator or cabinet doors. The contrast with a Soviet department store, such as GUM in Moscow—where the goods were behind the counters and one had to queue three times, once to select, once to pay and once to collect—could not have been more obvious. Not surprisingly, in certain exhibits (as with the Levi jeans), there was considerable attrition. The book exhibit was a key example: new supplies of books had to be flown in regularly. Even after the exhibition staff glued the spines down, the pages themselves disappeared.

Crowds of visitors watch a baking demonstration in the Jungle Gym.

Visitors throng the aisles of the Jungle Gym.

"Leaving the exhibition I carry with me an impression of glittering metal saucepans."

Soviet visitor

The "American experience" was reinforced by a wide variety of exhibits located outside the Dome and the Glass Pavilion. These included an exceptionally popular 360° Circarama film by Disney (seen in a small geodesic dome by 14,000 people a day), a continuous fashion show, the "Family of Man" photography exhibition, an American architectural exhibit, the Splitnik model house, a display of automobiles, boats and camping equipment, a beauty treatment and cosmetic kiosk, voting machines, a Pepsi-Cola refreshment kiosk, a children's playground, and so on.

One of the most popular outdoor exhibits was Edward Steichen's "Family of Man." Steichen, former Director of Photography at MoMA, had spent three years bringing together this collection of over 500 photographs, prior to its first showing in New York in 1955. Steichen believed that the portrayal of family life in all its diversity would bring unity to a world divided by politics and living under the threat of nuclear extinction: the concluding image of the show was of a hydrogen blast. "Family of Man" was famously attacked for its sentimentalism by Roland Barthes when it was shown in Paris in 1956. Barthes's view was, one now realizes, colored by the political status of postwar France, but there is a grain of truth in his view that the selection of work ignored large aspects of contemporary reality. Eric Sandeen's analysis of the use USIA made of the exhibition, while it does not mention the Moscow showing, implies that the "Family of Man" was seen by USIA as a form of soft propaganda, contrary to Steichen's intentions. This is a view that remains questionable, especially as Sandeen chooses to focus on the USIA's use of photographs of the exhibition, not on the exhibition itself and its impact.

A boy wearing a Komsomol scarf looking at kitchenware in the Jungle Gym.

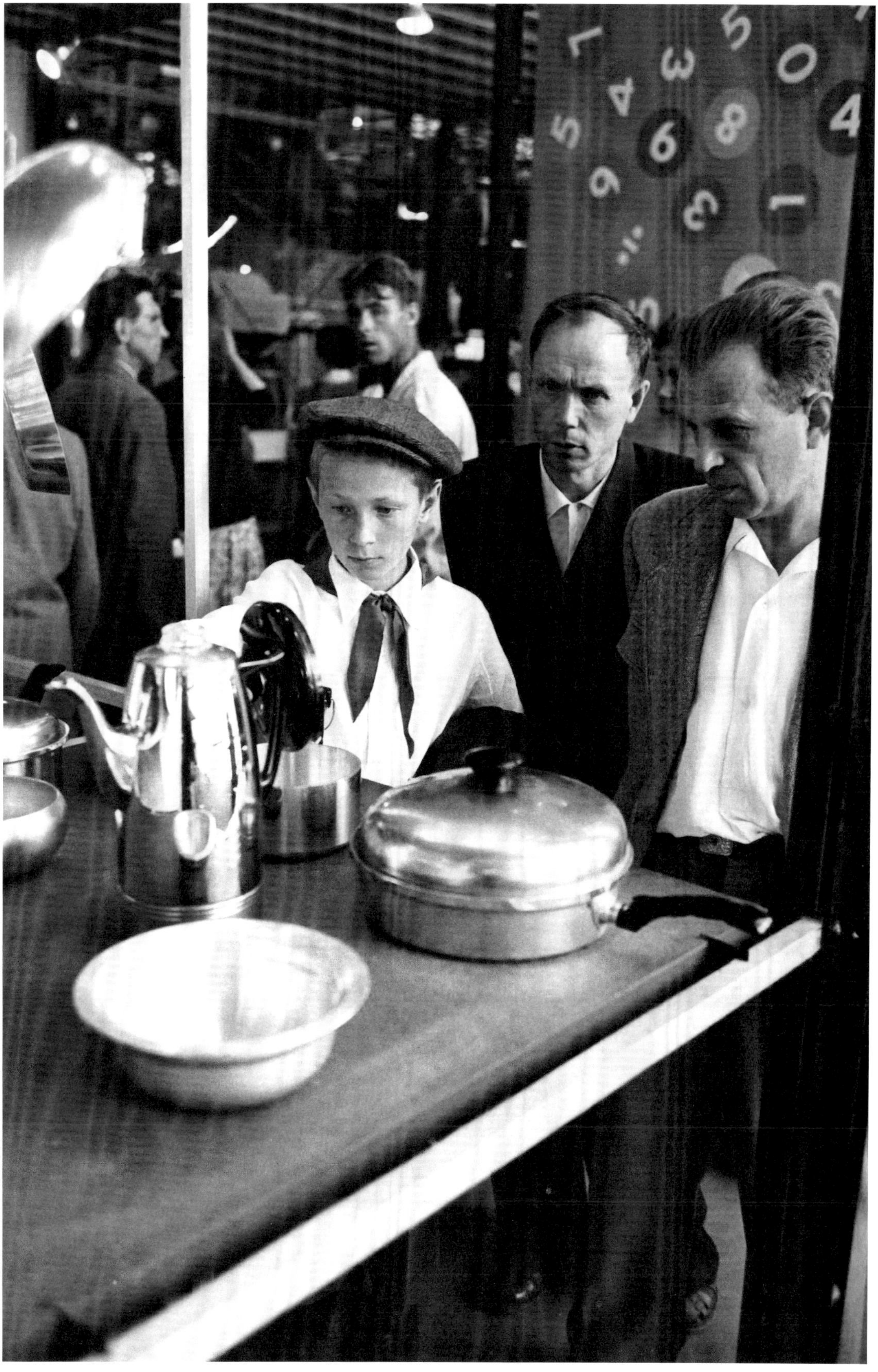

Peter Blake curated the architectural exhibition, using, as he had in Berlin, full-scale photographs of facades of buildings to convey the variety of contemporary American architecture.

For the beauty and make-up demonstrations, Helena Rubinstein herself joined the team on site to offer Russian ladies a Western look. The fashion show, too, attracted large numbers of visitors: the clothes displayed were both everyday dress and *haute couture,* both for men and women. Visitors often wept at the fashion show's concluding wedding scene, featuring a bride, groom and wedding party. Pepsi-Cola also had a kiosk: the drink met with Khrushchev's approval and so Pepsi was later enabled to export its cola to the USSR. The exhibition of the current year's automobiles fascinated Russian visitors, for whom the prospect of owning one of the limited range of cars available in the USSR was a daydream at best.

Posted throughout the American Exhibition site were seventy-five young Russian-speaking American guides, who explained the exhibits and answered the myriad of questions about life in the US that were continually asked by visitors.

When planning the visitor experience for the millions of Russians expected to attend the American Exhibition at the seven-acre Sokolniki Park site, the design team knew they could not structure a single storyline or a sequential traffic flow through the enormous site. Rather, they opted to provide multiple types of experiences in order to give individual visitors a choice of what to see.

As Irma Weinig puts it, for the Russians "the exhibition was a carnival: the brightest most colorful spot in Moscow. Usually deadly serious and law-abiding, living in a world of clearly-marked 'don'ts' and 'don't touch's,' they were free to follow their own bent at the exhibit. There were a thousand things to see, and no directions on what to do first. They dashed about, walking across grassy areas (in the city they not only would never do such a thing, but make sure no one else did either)."

Any attempt to analyze visitor responses must take into account the unexpectedness and variety of the event for visitors. As Weinig's article concludes: "If it was not the show of actual machines and technical processes which the Russians expected and wanted, it was perhaps something better, for it went beyond the machine to suggest the kind of life that an advanced technological society provides for its citizens. Someday Russia too may be less awed by the machine, more delighted by its product."

Avid readers looking at books,
magazines and newspapers
in the books exhibit: note the
gaps on the shelves.

Soviet women looking at stockings and shoes in the Jungle Gym.

A cooking demonstration in the television studio.

The baking demonstrations always attracted visitors.

Four views of the model apartment
designed by Lucia DeRespinis
and installed in the Jungle Gym.

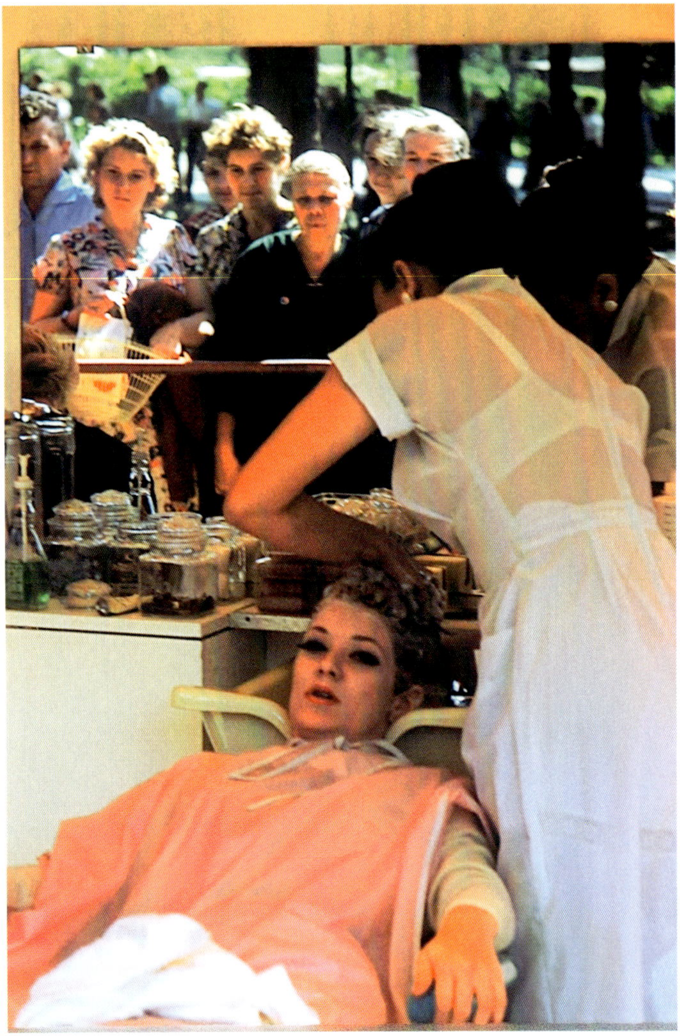

Marilyn Harris Bernard, top model for Helena Rubinstein, demonstrates make-up techniques.

A Coiffures Americana shampoo demonstration.

Artist's impression of the Beauty Kiosk.

A curious onlooker at the kiosk.

"The things I remember most were the autos, and the Pepsi-cola, and the booklets. The booklets were wonderful, such fine quality ... the colors, the printing. We had never seen anything like this before. The Polaroid camera to us was like magic. How was it possible to have an instant picture?"

A Soviet émigré to the US

Pepsi-Cola poster
for Moscow.

The Pepsi-Cola
refreshment kiosk.

Sailboats and motorboats in the man-made lake.

A Chrysler car under scrutiny from Russian visitors.

American dancing was a feature of the Fashion Show.

A Russian sable fur coat was a favorite with Russian visitors.

“It was very warm, that is, the way the people there were always with the smile. They were people who looked different, walked differently, they were like aliens!”

A Soviet émigré to the US

Haute couture fashions were professionally modeled.

> Menswear was also featured in the Fashion Show.

The entrance to the Architecture exhibition.

Visitors looking at the large scale photographs of traditional and contemporary architecture.

"Many, many pictures of people, beginning when they are very young and just meet each other ... I was so impressed, it was like something struck in my mind."

A Soviet émigré to the US

The entrance to the Family of Man photographic exhibit.

A pair of visitors looking at the "couples" section of the exhibit.

COLD ART

The art exhibition at Moscow, both of paintings within the Glass Pavilion and the sculptures shown outdoors, featured in part abstract art. Khrushchev was frankly dismissive of it: many Russians were astonished. To what extent, then, was abstract art used a political tool by US agencies during the Cold War? In 1974, the writer Eva Cockcroft published an article in *Art Forum,* entitled "Abstract Expression: Weapon of the Cold War." The thesis of the article was that the Museum of Modern Art in New York had collaborated with the CIA in the early 1950s to promote Abstract Expressionism as the vehicle of American cultural values and proof of the American hegemony of the arts. Nine years later a French academic, Serge Guilbaut, returned to the same theme with his book *How New York Stole the Idea of Modern Art.* In 1999, the English writer Frances Stonor Saunders revisited the same area in her wider study of CIA support for the arts in general, *Who Paid the Piper?* The catalyst for all these books was the revelation in 1967 that the CIA had been funding organizations such as the Congress for Cultural Freedom and the UK magazine *Encounter.*

"Abstract Expressionism" was a term developed by critics such as Clement Greenberg and Harold Rosenberg to describe the work of a group of postwar artists, and in particular Jackson Pollock. For these artists, according to the critics, it was the expression of the artist's relationship to the canvas and to pigment that was at the heart of artistic practice: the gesture was all. In the case of Pollock, these gestures included pouring, throwing and dripping paint onto the canvas: "action painting," to use Rosenberg's definition. The artists involved developed these various techniques in the mid-1940s: the critical discussion of their work began after the War. This new approach, according to Greenberg, divorced contemporary American art practice from its European roots and from any historical, moral, political or ideological context. Painting now existed for its own sake. Few would disagree that Abstract Expressionism and its later development, field painting, were largely American movements.

In Russia in 1922, as David Caute points out, the Association of the Artists of the Revolution dedicated themselves to "revolutionary realism," and exhibitions of modern work ceased. By 1925 many modern artists such as Chagall and Kandinsky had chosen to leave Russia, and in the early 1930s the process of State control of artistic subject matter intensified. "Formalism" regularly came under attack and Social Realism became the required order.

The style of Abstract Expressionism can certainly be seen as being at an extreme opposite from the style of social realist painting, preferred—or rather mandated—in the Soviet Union and its dependent countries. The historical models for Social Realism were the Russian nineteenth century "Wanderers," and any influence from later schools—from Impressionism onwards—was anathema. This was painting in the service of the State and the People, not the expression of the artist's personal vision or emotions. So one could make a case for Abstract Expressionism being used to "confront" Social Realism. But to what end? Abstract Expressionism was fiercely debated in America, in artistic circles and beyond. It was not well received in Europe either.

A further question is the chronology. Greenberg's enthusiasm for Pollock and others dates from the early 1940s, but it was not until a decade later that the Museum of Modern Art started buying or exhibiting work by such artists. The first American art to be seen in Russia was at the time of the American National Exhibition at Sokolniki Park in 1959, and even then abstract work formed only a part of a wider selection. The Museum of Modern Art had been sending exhibitions of modern and contemporary art to Europe from the early 1950s, often including Abstract Expressionists but never exclusively.

Part of the problem was that Abstract Expressionism—and a lot of related Modern art—was not only hotly discussed in the art world but outside it as well. The opponents of Modernism within the USA ranged from right-wing Congressmen to traditional art magazines such as *American Artist.* That Modernism was Communistic was a frequent allegation in the McCarthy years (certainly some artists, such as Ad Reinhardt, had been in contact with the prewar Communist Party of the USA). But the idea that abstract art was directed by Moscow, or that Abstract Expressionist paintings were coded maps of American defence locations was simple hysteria. Both the right wing in the US and the Communists in Moscow agreed that abstract artists were mentally deranged: for the Americans that meant they must be Communists, for the Russians that they could never be!

In January 1959, USIA commissioned a jury of four experts to select art works for the Exhibition. Among those chosen were "realist" paintings by John Sloan, Edward Hopper and Grant Wood, and "abstractionist" works by Arshile Gorky, Willem de Kooning, Mark Rothko and Jackson Pollock. The committee claimed that the works selected formed "the broadest, most balanced representation of recent American painting... so far shown abroad by our government."

When news of the jury's selections became known, Congressman George Dondero, a powerful McCarthyite, warned that the abstractionist paintings would "addle the brains of decent, innocent Americans." News of Dondero's unhappiness with the art show

> A Russian visitor contemplating an abstract sculpture of a woman.

was picked up by TASS, the official Soviet news agency: "Statements of the arch-reactionary congressman show once again the true worth of the fable about 'freedom' of artistic endeavor in the U.S."

Congressman Francis E. Walter, Chairman of the House Un-American Activities Committee, followed up on Dondero's rantings. (He would later accuse thirty-four of the artists represented in the art show of having had contact with subversive organizations or communist fronts.) Before Walter's committee could hold hearings on the show, however, USIA Director George V. Allen, had all 67 works of art chosen by the committee quietly shipped to Moscow. Eisenhower, an amateur painter who once said of the abstractionists, "Well, they have to eat," refused to withdraw any of the paintings. Edith G. Halpert, the first curator of the American show and fervid supporter of artistic freedom, tartly responded, "Well, some people don't think the President's paintings are so good either." At Eisenhower's suggestion, another 26 "traditional" paintings were later added to the Moscow art exhibit.

So it should not be surprising that the art show in Moscow became the single most controversial exhibit in the American Exhibition, a lightning rod for criticism from voices on both sides of the Iron Curtain. Victor Kemenov of the Academy of Arts of the USSR did not see "the relationship between the artist and the canvas" at the American show, but rather "canvasses ... with spots of dirty paints and ... freakish lines." In *Sovietskaya Kultura,* Kemenov declared that America's abstract artists were "like inmates in an insane asylum." Voicing the official party line, he wrote that the US valued abstract art, "for indeed it distracts the attention of the people from the hard problems that are created for them daily by life under the conditions of capitalist reality."

Kemenov's assertion that American abstract art was a tool of right-wing forces was but one assertion among many that abstract art was used by a US political cabal to promote a secret agenda. (In Washington, the very same "right-wing forces" that Kemenov accused of supporting abstract art had already mounted a fierce assault on it!)

Soviet visitors of all ages attended
the exhibition.

VISITORS AND GUIDES

After the Exhibition closed, a USIA report estimated that 2,700,000 Russians had visited the exhibition during its six-week run. This number was arrived at by extrapolating from head counts at the entrances. Ticket sales suggest a lower figure, but it soon became clear that people were getting in without tickets—not that the Americans minded. Ticket sales were initially controlled by Soviet trade unions, so, while there are no statistics available on which groups were favored for tickets, there was good reason to suspect tickets were given to loyal Party members.

Moscow was both the administrative and political capital of the Soviet Union and a major industrial base, with a population of over five million in 1959, having grown by one million since 1939. While this number would include the Soviet *nomenklatura* (senior Party officials and top bureaucrats) and industrial managers, the large majority of Moscow's population were workers in factories, service industries and offices, who often lived in crowded apartment blocks, at worst—and not so rarely—with one family to a room sharing cooking and washing facilities. The working week was 46 hours, and many Muscovites had to have two jobs in order to earn enough to live on. (The Moscow comment on Khrushchev's

The guides meet the President at the White House before going to Moscow.

Znachki—souvenir buttons—were prized visitor trophies.

promise to reduce the working week to 35 hours was: "Good, that makes more time on the second job.")

The majority of visitors to the Exhibition would therefore have been drawn from this urban proletariat, a minority from the *nomenklatura* and their sphere of influence, and a further minority from beyond the orbit of the Moscow *oblast* (city area). There is evidence from the guides that the visitors in the opening weeks came from the *nomenklatura:* this was seen in their style of dress and language, and the fact that many wore medals or party badges; tickets were initially distributed to favored groups, and only later went on general sale. Once tickets were available more widely, the mix of visitors changed to reflect the whole of Muscovite society.

Travel to Moscow from other cities was at that time regulated by an internal visa system, so any visitors from outside the city would have to obtain permission to visit Moscow. Many, including students and those with relatives in Moscow, were given permission to visit. There were others, for example, residents of the Baltic States forcibly made part of Russia at the end of World War II, who came on their own initiative. At times, comments by the guides distinguish between Muscovites and others, but the main distinction their comments make is between "agitators" and ordinary citizens.

Faces in the crowd at the Fashion Show.

The Komsomol, or Young Communists League, had been training its members for "agitation" almost since the moment the Exhibition was announced. In June 1959, the Moscow Bureau of the CPSU (Communist Party of the Soviet Union), as Hixson points out, involved CPSU and Komsomol officials, trade union leaders, industrial managers and newspaper editors in a campaign of "mass political, cultural and educational work for the summer period," that is, for the duration of the Exhibition. Their opposition took a number of different forms, including promoting the Soviets' own exhibition immediately adjacent to the American one, and, of course, making tickets to the American Exhibition difficult to get. A confidential USIA report presented in September 1959 details actions such as putting up posters for the exhibition, but removing the part that explained where tickets were available or the closing date. (The same report suggests ticket seekers were made to wait in line twice, once for chits for tickets, then for the tickets themselves, though in fact this only mirrored standard practice in Russian shops.) Once inside the Exhibition, targeting the guides with rude and aggressive questions about unemployment or lynching in the US was a standard political procedure. Before evaluating the effectiveness of this heckling one should understand who the guides were and how they managed such questions.

The 75 guides at the Exhibition were between the ages of twenty-five and thirty, and included both native Russian speakers and Russian language graduates. 25 were women, 48 were men, and among the group were 4 African-Americans. In the minds of USIA planners, the guides were "live exhibits" who could personally interact with Soviet visitors, most of whom had never met an American. The guides were selected after an interview and language test: they then embarked on a training program in New York

American guide Tania Sochurek greets visitors at the "Splitnik".

Demonstrating a mechanical sofa-bed in the Jungle Gym.

Talking to guides, whether in the display of sewing machines, the Family of Man exhibit or the outdoor exhibits, was a key experience for Soviet visitors.

before going to Moscow. The significance of their role is reflected in their June 15, 1959 meeting at the White House with President Eisenhower before the start of their training. (Eisenhower encouraged them to smile and "not to brag to the Russians").

The guides had their travel, board and lodging paid by USIA and were paid $16 each day while in Moscow, but the opportunity provided was beyond remuneration for those planning to use their language skills as a career (nearly eight hundred candidates applied). Some even arranged to travel around the Soviet Union before and after the exhibition: one was John R. Thomas, who compiled a lengthy report on his time at the Exhibition and his meetings with over seventy-five individual Soviet citizens in Moscow, and on a trip to Latvia. This report was sent to USIA from the Rand Corporation in Santa Monica, California. His report echoes the comments of other guides.

Thomas was stationed at the voting machines, meant to symbolize American democracy in action, which were used to record visitors' reactions to the exhibits. When Khrushchev visited the Exhibition with Nixon, the Russian leader pointedly ignored the voting machines, cutting off his Presidium colleague Frol Kozlov, who had started talking to Nixon about them. The machines were clearly a target for "activists and other zealous defenders of the Soviet regime," as Thomas calls them. Particularly after the rush of initial official visitors was over, he found that these hecklers would at times be shouted down by other Russians in the crowd, and that the best way to deal with repeated allegations on subjects such as race and poverty was not to get into an ideological debate, but admit to faults in the American system while stressing that progress was being made. Much of his report deals with political issues, but he also adds that he was asked endless questions about the standard of living in the USA, and about his own life and circumstances. Many of the guides at first found this inquisitiveness rude, but soon realized that asking someone how much they earned—or their father earned—was as welcome a question in Russia as it was impolite in the USA.

The Russians' insatiable curiosity about Americans—and all things American—is understandable given the demonization of the USA in the Soviet press, the lack of contact with foreigners in general, and the boredom of everyday life under the Soviet system. In the opinion of the guides, most Soviet citizens were skeptical about much of what they had been told by the State, though they showed this through silence rather than overt responses. Interest in the objects exhibited was as much a pretext for talking to the guides as anything else. Many visitors followed the official line that, as guide Ellen Propper put it, "the exhibition was filled with the superfluities of life which were inconsequential for a state constructing Communism. However, beneath this was a real fascination with the objects exhibited and a genuine, if fleeting, desire for them." Many of the questions were about the cost of goods, about salaries, and about schooling and health care. Their knowledge of America, according to guide Edith Rogovin, "ranges from complete misinformation to amazingly accurate facts." Many believed, Dan Slobin pointed out, that "atheism is against the law in America." Several of the guides sensed the Russians' mixture of "pride, but obvious backwardness," in Joyce Klain's words, and John de Luca notes that "there is a strange blend of cynicism and optimism about the prospect of satisfying commercial and intellectual needs among young people." The Russians considered themselves misunderstood in the West: "Tell Americans the truth that we're simple Russian people who want peace and friendship," as Joyce Klain put it. The guides also realized that the average American's understanding of Russia was defective as well: as Chas Gribble pointed out, "The worst false impression is that most Russians are terribly dissatisfied and will easily revolt against Communism."

The guides were among the most articulate ambassadors at the exhibition, but they were not the only Americans present; there was a team of forty-five for the fashion show, technical crews for the television studio, the beauticians, specialists for the automobile exhibition, and the bakers in the Jungle Gym, many of whom were Russian speakers. In the opinion of many, it was the presence of these Russian-speaking Americans who engaged in one-on-one conversations with the Soviet visitors who made the optimum impact of the Exhibition in Moscow.

“What did we Americans see? What was below the surface? Hunger to know—and to see—and to believe.”

George Malko, Associate Producer, NBC News

Visitors clutching guidebooks and leaflets listen to a guide in the Jungle Gym.

Visitors hoping for free samples from the Beauty Kiosk.

Waiting for answers from the RAMAC computer in the Dome.

SPUTNIKS OVER BROADWAY

The Soviet Exhibition in New York, which was the USSR's reciprocal national exhibition, faced fewer logistical problems than did the American Exhibition in Moscow. It was held in the world-class exhibition facility, the Coliseum in New York City; more importantly, the Soviets brought their own world-class exhibit, the Sputnik.

Sputnik 1 was the first man-made satellite to circle the Earth. Launched by the Soviet Union in October 1957, it created consternation in the USA, whose presumed leadership in space technology and nuclear weaponry was immediately undermined. The fact that the next two American attempts to launch orbital rockets failed disastrously only underlined this uneasiness. There is little doubt that the launch of Sputnik, a ball full of liquid nitrogen and a radio transmitter, had an astonishing effect on US opinions and policy: within a year NASA was created and the Space Race which was to culminate in the moon landings had begun.

The Soviet Exhibition, therefore, touted Soviet strengths while exposing their insecurities, thus mirroring the schizoid combination of superiority and inferiority complexes that defined the Soviet attitude to the outside world—and, as some would say, Nikita Khrushchev's own often polarized behavior, veering between bombast and concession. Sputniks and statistics would well describe the Russian exhibition in New York. It was billed as "an exhibition of Soviet achievements in the spheres of science, engineering and culture" where "visitors would have the opportunity to familiarize themselves with the Soviet Union's state structure, educational system, public health services, culture and way of life." The 170,000 square foot exhibition included three Sputnik satellites, a model rocket and a working model of an iron and steel making complex among some ten thousand exhibits. In the cultural sphere there was an art exhibition: the press officer for the event, Enver Mamedov, promised an art exhibition "for the millions, not for the aesthetic few." No abstract art there! There was a fashion show, a model apartment, and English-speaking Russian guides (who met their American counterparts in Moscow after the New York show finished). They seem to have been as exhausted by American questions as their counterparts were by Russian ones, one American guide recalled.

The Soviet Exhibition in New York drew fewer visitors than did the American Exhibition in Moscow, and was not met by the barrage of press criticism the Americans incurred in the Soviet Union. Indeed, US press coverage had much that was complimentary to say in the run-up to the show, but little to say about it after it opened. Columnist Bob Considine wrote: "The Russians (have)

Inside the Soviet Exhibition in New York.

The Soviet "Achievements of the People's Economy" exhibition in Moscow competed with the US exhibition at Sokolniki.

Cover of the handbook to the Soviet Exhibition.

put together an impressive exhibition, better designed and not so gooey on the propaganda side as their gigantic barn of goods at the Brussels World's Fair last year." In response to the Soviet exhibit featuring Laika, the first dog in space, Eisenhower reportedly quipped, "Well, we put a mouse up."

The difference in content between the two exhibitions, in Moscow and New York, lay predominantly in the Soviet emphasis on their technological achievements, especially in space, which was wholly absent from the American show, perhaps because America's space program was experiencing problems ("our boys always botch it and our rockets always blow up," as Tom Wolfe said).

LIFE AFTER SOKOLNIKI

"Our people grumbled for more nylons and washing machines"—the line is spoken by that consummate actor Peter Bull, playing the role of Soviet ambassador Desadeski (a wonderfully evocative name) in Stanley Kubrick's 1960 film *Dr. Strangelove.* The film is a nuclear black comedy with Peter Sellers playing multiple roles (the President of the USA, an RAF liaison officer, and the eponymous doctor): in it, the Soviets resolve the problem of funding the arms race by building a doomsday machine. It was a funny and unnerving film at the time, and still is.

Making them grumble for nylons and washing machines was one of the aims of the Sokolniki Park event. The other aim was to break the image of Americans as capitalist warmongers and replace it with something more realistic and accurate, stressing the standard of living and freedom of choice an average American could enjoy under a democratic system. How far were these aims achieved?

Ralph K. White, a USIA officer (who would later become famous for his studies on the psychological causes of war) visited the Exhibition in Moscow specifically to gauge visitors' reactions. On September 28, 1959, three weeks after the Exhibition closed, he issued an initial evaluation of its effectiveness. Stressing that the report was a preliminary one, he noted that "success [of the Exhibition] was moderate and somewhat equivocal." He begins by dismissing the results from the voting machines and the comment books, though positive, as being unrepresentative. He relies much more strongly on the feedback from the guides, the comments overheard by them and others in the crowds of visitors.

While agreeing with the official USIA assessments and American press reviews that the exhibition was successful, White challenges the degree of success. He identifies a number of areas in which the exhibition fell short. Firstly, he notes that Soviet visitors had exaggerated, unrealistic expectations of what they were going to see. Secondly, that "visitors were genuinely disappointed in the scarcity of American technology at the exhibition." He noted: "The demand was not for heavy industry" but for "wonderment at mechanical parts busily in motion." Thirdly, he identifies the confusion caused by a general lack of signage and visual information to guide the visitors through the exhibition site.

Fourthly, he cites the lack of sufficient *znachki* (souvenir buttons), leaflets, plastic, and other memorabilia which the visitors avidly sought as a token of their attendance. "Perhaps," he writes, "all of them were badges of prestige, since having been to the Exhibit was itself, at least during the first few weeks, a source of prestige. Perhaps they had symbolic value as tangible signs of personal contact with a great foreign country ... Perhaps the chief thing they symbolized was contact with the great outside world ... But whatever the psychological reasons may have been, the trophies clearly had great symbolic value."

White cites in fifth place the failure to display American luxury sufficiently, which he identifies as the counterpart to the Soviets' craving to see the very best America had to show, in technology as well as consumer goods. Finally, he suggests—admittedly with little evidence to support it—that the exhibition had too little about "the life and spirit of America." In his personal opinion, "the preoccupation with consumer goods was somewhat too great ... with too little on actual living in America ... or on the essentials of democracy ..."

In his opinion, the "Family of Man" was "first in total impact," though he noted that the automobile show, color television studio and the Circarama film were also strongly favored, according to votes and comments. The popularity of these easily accessible exhibits is not hard to understand, just as the unpopularity of the abstract art was inevitable in a country that had shut the door on abstraction thirty years before. White's own comment on the Eameses' film was to wish it had been slower. This is a feeling many Muscovites might have reciprocated, as the pace of the film must have seemed extraordinarily fast to them. This may explain why the Eameses' film is mentioned so little in the visitors' books and voting machine comments. It is ironic, nonetheless, that what is now seen as a key event in the Eameses' career as filmmakers was not perhaps wholly appreciated by its intended audience. (Writing in 2005, Jeffrey L. Meikle claims that "by all accounts 'Glimpses of the USA' was the hit of the exhibit." Perhaps it was to Americans who saw it, but the truth may be more complex. That it can be seen as a turning point in the Eameses' career is not so doubtful.)

Was this gap in communication, perhaps through over-sophistication, something seen more widely in the Moscow exhibition? Almost certainly not: there may have been misunderstandings on the part of the visitors, such as those who refused to believe that the Splitnik was a house for a single family as it had two bathrooms. The Russians' enthusiasm for the American effort—and their endless curiosity—are also indicators of a certain measure of success. It was in America's interest to undemonize the image of the USA, to put forward an image of the country as proud as well as peaceful. If in the process the Soviet people began to listen less avidly to what their own government and Party officials were telling them, well, they were not that gullible—at least in Moscow—to start with. The way the guides saw visitors interacting with agitators on the US's behalf is clear enough proof of Muscovite scepticism about official propaganda.

The problem of analyzing the success of the Moscow Exhibition is that there was nothing to compare it with. Unlike World's Fairs, where visitors could compare pavilions from competing countries, in Moscow, the US had no competition other than Soviet preconceptions about the United States, and in this regard, the Americans had their work cut out for them. If the ordinary citizens of the USSR knew only what the Soviet government told them about America, US citizens' knowledge of the Soviet Union was scarcely more informed. Tourism between the two countries was almost non-existent, trade extremely limited, academic and cultural contacts were heavily monitored on both sides. If the expectations on the two sides differed, in terms of what the Americans thought the Russians wanted and in terms of what the Russians thought they were going to get, that was almost inevitable.

There were Russians who believed then (and who believe now) that the American Exhibition was a Trojan horse bearing subversive ideas, which were the equivalent of "poisonous blankets": this linked the US attempt to seduce the Soviets with chrome and consumerism to allegations that the US cavalry had given Native Americans blankets from cholera victims. As Vladimir Papernÿi writes: "Instead of smallpox or cholera, (visitors) were infected with poisonous ideology ... (it) worked brilliantly."

This evaluation notwithstanding, there were Americans at the time who felt that the consumerist vision was being overplayed, and that a profounder picture of American beliefs and values should have been painted. But part of the US-Soviet agreement specified "no propaganda," and that clause would have implicitly excluded presentation of issues of concern to the Soviet government, such as "religion," "democracy," and "individual freedom." At a more practical level, given the average Soviet citizen's limited knowledge about America, how could the whole of a culture be presented in an exhibition a visitor would attend for at most a day? (For that matter, how many average Americans themselves had a deep understanding of the diversity of American culture?)

Such a viewpoint ignores the question of what the Americans sought to achieve: unlike "Unfinished Business" at Brussels, in Moscow problems were deliberately not put on display—though these were often addressed by the guides. Nor was the content of the exhibition dishonest in any way. The Americans sought to put forward the best image of America, just as the Russians did at Columbus Circle.

In 2003, David Caute noted that the 1959 Exhibition was "generally a success," though it is not clear whether he means in terms of US and Russian perceptions of having achieved their own objectives, or as an example of improved relations attributable to the Cold War thaw. What can be learned from the American National Exhibition in Moscow from a USIA perspective is that it confirmed the success of a method of presentation, combining where possible impressive architecture with varied content, and, above all, human contact.

LAUTMAN IN MOSCOW

In 1959 the Kaiser Company asked the architectural photographer Robert C. Lautman to go to Moscow and photograph the Dome for them. Lautman had served as a combat photographer in the Pacific in World War II. On his return he was apprenticed to an architectural photographer—and, as his commendation for an AIA Gold Medal in 2007 puts it, "was charmed by his subject." He opened his own practice in Washington D.C. in 1948 and soon became the photographer of choice for many of the modernist architects in the region and beyond.

In Moscow, as well as carrying out his commission, he turned his camera on the crowds of visitors. Some of these images conclude this section—they are part of the archive he donated to the National Building Museum in Washington D.C., in 2007.

STERN
LEONARD
GLENN GOULD
KOSTELANETZ
FLOWER DRUM SONG
CARL SANDBURG
MONK

"It really was not worthwhile to bring your 'sculptures' to Moscow. They cause laughter and puzzle the spectators. Your automobiles are excellent. It is hard to understand how a country which builds such automobiles can show such 'art'."

Soviet Visitor

"About the exhibition, one can say in short: Brilliant, deep, exciting."

Soviet visitor

"If there was anything the Russians wanted to believe, it was that we were telling the truth—our truth—and they had such incredible experience of sweeping past propaganda and trying to find the heart of it."

George Malko, Associate Producer, NBC News

"The exhibition gives an impression that America is more interested in looking after comfort and amusement rather than the spiritual enrichment of man."

Soviet visitor

It Pays for itse
Savings!
the Modern SIDING

“This is what I remember about Moscow in 1959: Despite the things we didn’t like about the Soviet Union, we met people who were very much like us—brought up over there, but who had read the same books and who had the same human instincts. And that was all to the good.”

George Feifer, American guide

TRAVELING HOPEFULLY

AMERICAN TRAVELING EXHIBITIONS USSR 1961–1965

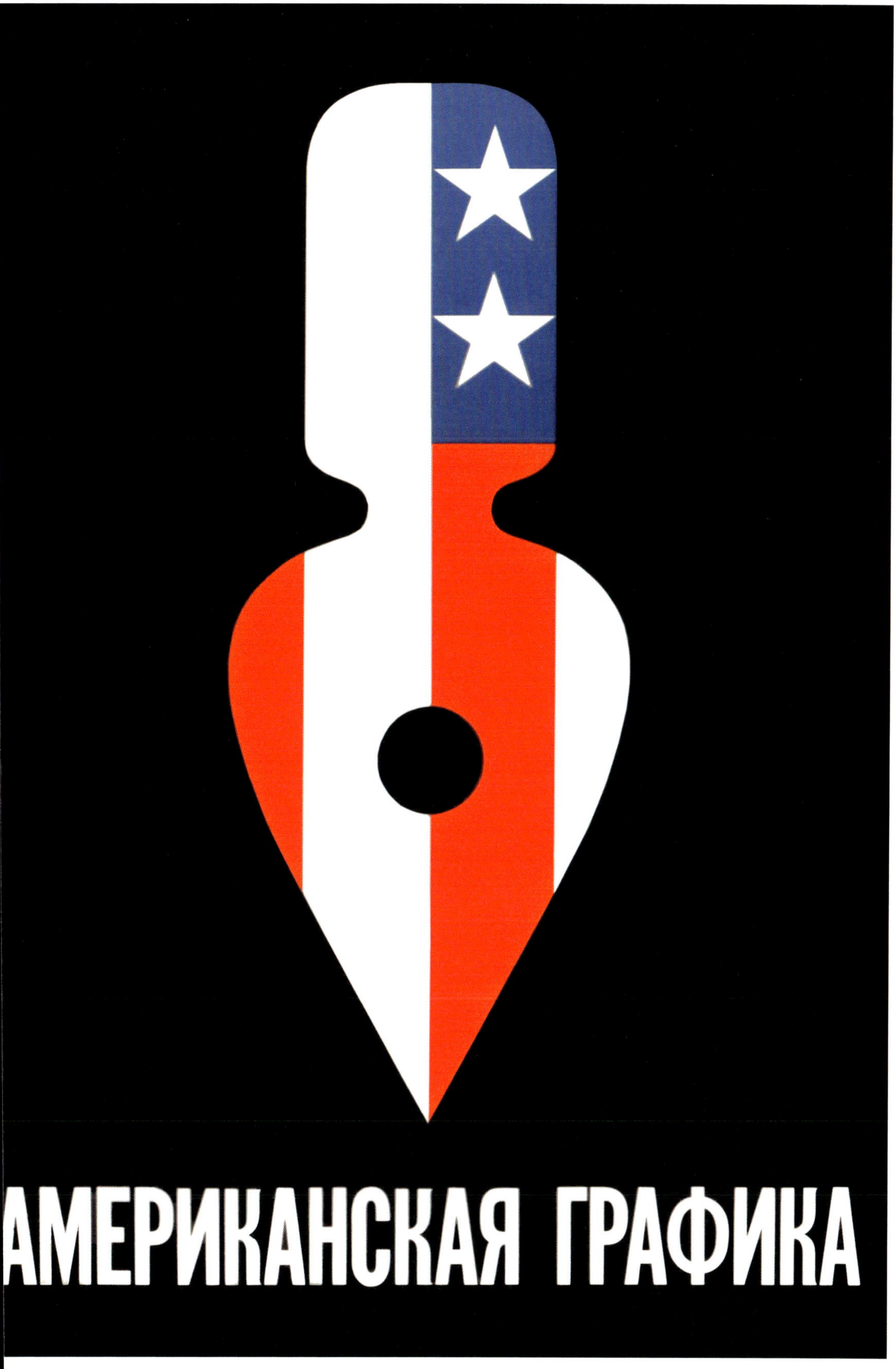
АМЕРИКАНСКАЯ ГРАФИКА

It's snowing in Moscow on a December morning, but people don't mind the snow there. They don't mind waiting a while, either: queuing is an endemic activity in 1960s Russia. What is not surprising is what they are queuing to see, an exhibition of graphics. What is surprising is that the work in question is American, and that a group of American guides, young, Russian-speaking Americans, is waiting to welcome them and discuss not only the work exhibited, but what life is like in the United States. That was something worth waiting more than an hour for, in line, in the snow.

But why was this happening at all? The early 1960s were the brinkmanship era of the Cold War, with Cuba soon to be the high tide mark of military confrontation. The 1958 Cultural Agreement, and the exchange of national exhibitions that followed in 1959, seemingly paved the way for clearer understandings between the USA and the USSR; and the visit of Premier Khrushchev and his wife to the USA in the autumn of 1959 seems to have led to a closer personal understanding between Eisenhower and the Russian leader, in particular after their meeting in the secluded confines of Camp David. There seemed to be a real possibility that the planned 1960 Paris Summit could lead to the signing of a treaty banning the testing of nuclear weapons, the first step in a slowdown of the arms race. This would be a triumph for both leaders, confirming Eisenhower's chance for a peace policy and Khrushchev's principle of peaceful co-existence.

However, this optimism literally fell out of the sky when Major Gary Powers' U2 spyplane was shot down in May 1960. The U2 spy flights had been instigated by the CIA in order to obtain photographic information on Russian military establishments. Eisenhower perhaps let others overrule his better judgement. His judgement failed again when he initially denied the existence of the flights. Even at the Paris Summit he refused to apologize, giving Khrushchev no option but to walk out. Khrushchev's opinion of the Americans returned to an uncompromising one: Washington was dominated by the hawks and the hardliners, who had no respect at all for the Soviet Union and its achievements.

However, the U2 incident did not put a stop to the series of smaller exhibitions that had been agreed on in 1958. The year 1961 marked the begining of an exchange of touring exhibitions between the US and

< The poster "American Graphics" was designed by Thomas Geismar.

the USSR. Soviet exhibits that toured the USA included themes such as medicine, technical books, graphic arts, public health, the conquest of space and "Soviet Siberia Today." Many American exhibitions, after touring the Soviet Union, were shown in other countries in the Soviet bloc, including Poland, Hungary, Czechoslovakia, Bulgaria, Yugoslavia and Romania. The first of these traveling shows to reach the USSR was "Plastics USA," curated by Jack Masey with Latham Tyler Jensen and others. It opened in Kiev in the Ukraine, and went on to Moscow, and Tbilisi in Georgia. A second exhibition in 1961, "Transportation USA," designed by George Nelson, visited Volgograd in the Caucasus (a city which until that year had been named Stalingrad) and Kharkov in the Ukraine. In 1962 the third touring exhibition, "Medicine USA," was shown in Moscow, Kiev and Leningrad. In 1963, the same three cities were the destination of an exhibition entitled "Technical Books USA." (Both the medical and technical books exhibitions were designed by Masey.)

One of the most popular American touring exhibitions, "Graphic Arts USA," was designed by Chermayeff and Geismar and was shown in Alma Alta in Kazakhstan in 1963. This exhibition then went on the next year to Moscow, Yerevan in Armenia, and Leningrad. A further new exhibition, "Communications USA" designed by Jay Doblin, was shown in Leningrad and Kiev, before going to Moscow in 1965. This exhibit was followed in the same year by "Architecture USA," shown in Leningrad, Minsk and Moscow: this exhibition was curated and designed by Arthur Drexler of the Museum of Modern Art, New York.

In addition, there were exhibits on hand tools, industrial design and education—topics in which the Soviets had expressed interest; these toured the Soviet Union and crisscrossed the Soviet bloc through the 1960s. As Eugene "Rocky" Staples, manager of the Plastics exhibit, put it, "The American side wanted our exhibits, even if narrowly focused by topic, to yield a broad, not a limited view, of American life."

The exhibitions were displayed in existing buildings and used free-standing panel systems that could be adapted to the constraints of the different spaces. Each exhibition would run for about a month in each location, with a month's transportation and installation time between each showing.

> Visitors wait in the snow to see the American Graphics Arts exhibition in Moscow.

The major USIA touring exhibitions in the Soviet Union were staffed by teams of Russian-speaking American guides. The guides were not only there to explain the work on display, but also to answer questions asked by the visitors about life in America. For many Soviet citizens it was their first opportunity to meet a real American, as opposed to the stereotypes presented by Soviet media.

Most of the guides were university graduates, men and women between the ages of twenty-five and thirty-five. Shepherding these people around during a six-month exhibition tour could sometimes bring its own problems. Prior to departing, the guides underwent a briefing in Washington about what to expect during their stay in the Soviet Union. They were warned to be especially aware of the danger of sexual entrapment. On occasion, however, a young guide would forget this advice, only to find him- or herself in a compromising situation: the American Embassy in Moscow was not unaccustomed to securing a place on the next flight back to the United States for a hapless, love-struck guide.

Visitors who talked to guides must have known that there would be others in the crowd monitoring them: what must have surprised them was that no one was monitoring the guides, who were able to talk openly about the United States and speak frankly about its problems. This personal contact between guides and Soviet citizens was one of the key successes of the touring exhibitions.

Queues to visit the graphics exhibition in Alma Ata.

President Kennedy with the President of the Lithographers Union.

Guides for the Graphics Arts USA exhibition at the White House after a meeting with the President.

GRAPHIC ARTS

The Graphics Arts exhibition was designed by Chermayeff and Geismar and began its tour in Kazakstan in 1963, before being shown in Moscow and two other cities in 1964. In 1965, the exhibition was seen at three locations in Romania.

The exhibition included a workshop as well as a display of prints. A printmaker with a well-equipped studio was part of the tour, and would demonstrate printing techniques. The American illustrator Norman Rockwell visited the exhibition and made drawings of young visitors. The leader of the Lithographers Union had a portrait made of President John F. Kennedy which was featured at the entrance to the exhibit: after Kennedy's assassination in 1963 a portrait of Lyndon Johnson, created in the Soviet Union by one of the guides, who was a graphic artist, was added.

While the graphics exhibited had no overt political content, the contrast between the diversity of American design styles and the codified imagery of the Soviet state would not be lost on visitors. Among the works shown were posters, paperback and album covers, logos, packaging, advertising materials, illustrations and artworks.

Russian visitors at the
Graphic Arts USA exhibition.

Russian visitors at the Graphic Arts
exhibition.

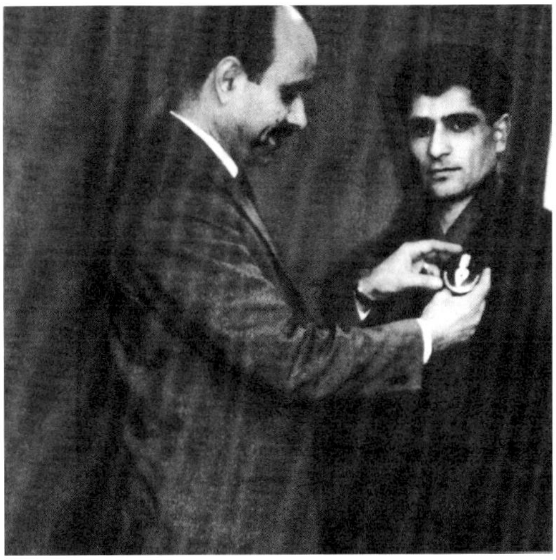

American guide welcomes the millionth visitor in Yerevan, Armenia.

Norman Rockwell drawing portraits of visitors.

The printmaking studio in action.

The exhibits were mounted
on a panel system.

The exhibition brought together logos, posters, commercial work and illustration.

SAVE A PILE!
FALCON
SEE YOUR NEIGHBORHOOD FORD DEALER

Рудолф де Харак. Обложка праздничного выпуска журнала графического искусства.

Print Magazine © Kaye-Cadel Publishing Corp.

Examples of American graphic arts.

ХУДОЖЕСТВЕННОЕ ОФОРМЛЕНИЕ

"We were interested in expressing ideas through color and shape and typography as opposed to just technique and skill."

Thomas Geismar, Designer

The stylistic variety of American work differed from the traditional approach of Soviet graphic designers.

ПОБЕДА
КОММУНИЗМА
НЕИЗБЕЖНА !

PLASTICS USA

Plastics USA was the first of the touring exhibitions in the Soviet Union. Just as the American National Exhibition in Moscow had focussed on the products of industry rather than industry itself, so here too the emphasis was on the variety and adaptability of plastics for use in automotive design, medicine, clothing, home furnishings and housewares.

Russian visitors exploring the wonderful new world of plastics and its myriad applications.

"We are supposed to be talking about plastics, but the people are much more interested in learning about us and so I talk about life in America, about myself, about literature, about my impressions of their country."

Sylvia Juran, American guide

TRANSPORTATION USA

Transportation USA, designed by George Nelson, was the second exhibition to travel in the Soviet Union. It featured a Cessna Skyhawk airplane, a Ford Thunderbird hardtop car, a 14-foot model of a Boeing 707 passenger aircraft, a scale model of a concept car on a circular track, as well as information on public transport systems. As with other exhibitions, the emphasis again was on the range of choice available to the American traveler.

Cars and planes excite the visitors.

TECHNICAL BOOKS USA

This exhibit contained a collection of 7,000 representative technical books and reference materials showing the variety of technical publishing in the US. The exhibit came closest to the Soviets' official desire to learn more about American technology. The books shown included scientific works as well as titles used in secondary and university education.

Given the high value set on orthodox academic learning in Soviet culture, this exhibition was a considerable success: as with the books section of the Moscow exhibition in 1959, replacement stocks were regularly needed.

Not content just to look,
visitors settled down to read.

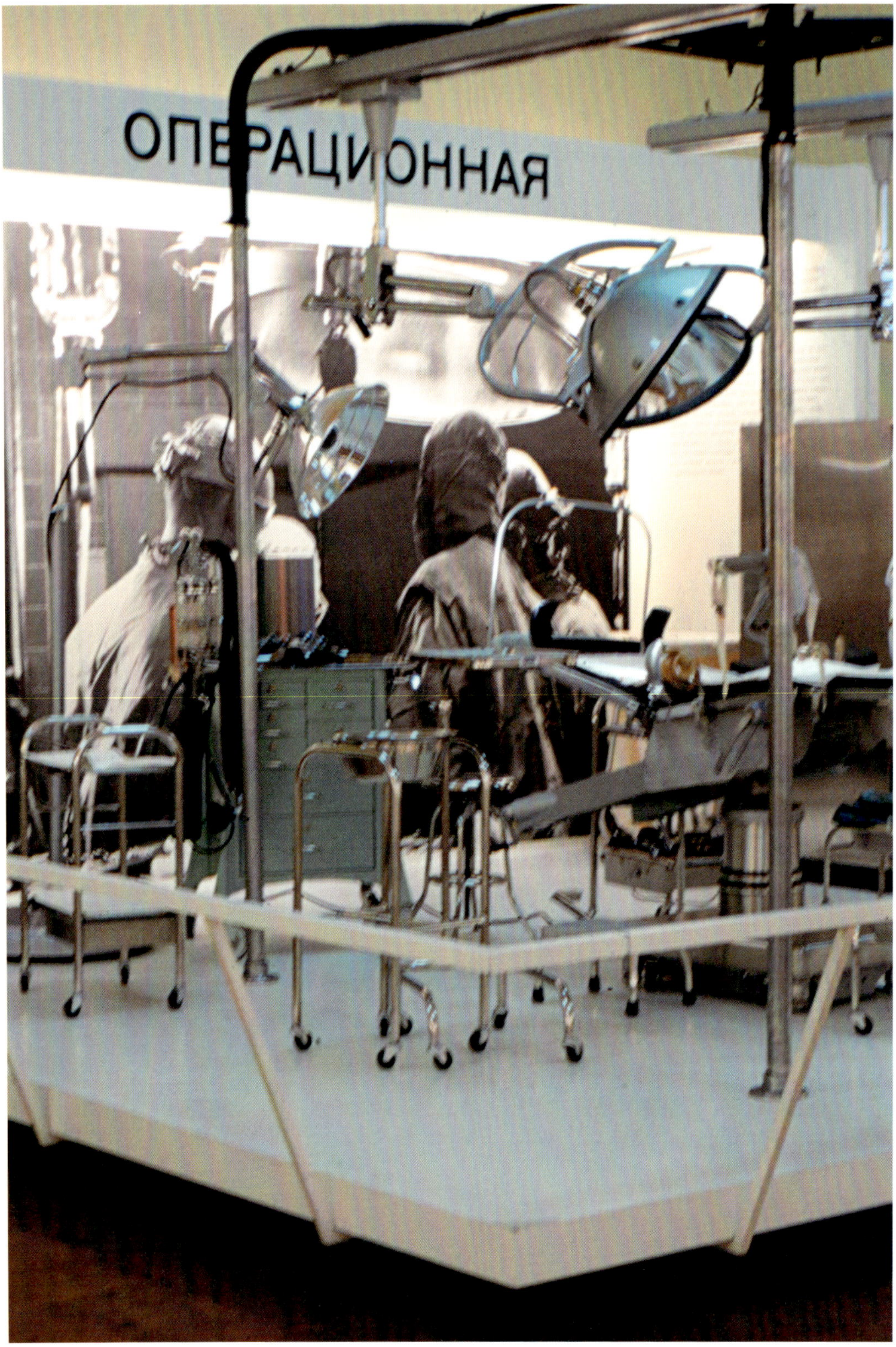

A full scale operating theater.

MEDICINE USA

The 7,000 square foot exhibit included a full-scale operating theater, a section of a typical drugstore, a dentist's office and a variety of medical equipment including a heart-lung machine, an atomic scanner, a neonatal room and a representative assortment of medical disposables and instruments.

A replica drugstore.

MONTREAL MAGNIFIQUE

UNITED STATES PAVILION
CANADIAN WORLD
EXHIBITION
MONTREAL, CANADA
1967

USA

In the nine years that separate the two official Universal Expositions, those of Brussels in 1958 and Montreal in 1967, a great deal had changed in the Cold War. The brief understanding that opened between Khrushchev and Eisenhower after Khrushchev's visit to the USA was abruptly terminated by the U2 incident. When John F. Kennedy was sworn in as President in January 1961, he rapidly arranged a summit meeting with Khrushchev in Vienna that summer. While there was some agreement at the meeting over the question of Laos, Khrushchev's incessant return to the theme of Berlin led to no positive results, save that the Soviet Premier was convinced that Kennedy was the weak pawn of militarists and industrialists, while the American president was convinced that the Russian leader was dominated by hardliners in the Politburo. Neither of these assumptions was correct: it was a poor start. Khrushchev, frustrated by the Berlin issue and the endless flight to the west of East Germans, ordered the building of the Berlin Wall in August, constructing one of the classic symbols of the Cold War.

Just as Brussels had been held under the shadow of Sputnik, so Gagarin's successful space flight in 1961 influenced the first year of the Kennedy presidency. Despite the visible failures of American rocketry, Kennedy announced that the Americans would put a man on the moon within a decade: the space race joined the arms race as a further dimension of the Cold War.

Fidel Castro's seizure of power in Cuba created another area of potential confrontation. While more Marxist than Stalinist, the Castro regime, under an American trade embargo, looked to the Soviet Union for support. The disastrous Bay of Pigs invasion of Cuba by forces from the USA consolidated Castro's power and led the Soviets to their biggest gamble of the Cold War: sending military personnel, and tactical and medium range nuclear weapons and missiles to Cuba in 1962. The October stand-off between the superpowers ended with a *quid pro quo:* Soviet withdrawal, against a promise from the USA to withdraw its missiles from Turkey and a secret understanding never to attack Cuba.

A new agreement between Kennedy and Khrushchev led to the signing of a Partial Test Ban Treaty in August 1963, and after Kennedy's assassination in November of that year, President Lyndon B. Johnson

< Visitors at the entrance to the Pavilion.

and Premier Khrushchev simultaneously announced plans to scale back the production of nuclear weapons materials.

But as one door closed, another opened. If there was greater macro-stability, new problems arose elsewhere. Though the French had attempted to retain their colonies in Indo-China, Vietnam, Laos and Cambodia after the Second World War, an insurgency by the Viet Minh became a full-scale conflict, and the French were defeated at the battle of Dien Bien Phu in the spring of 1954. The Geneva Accords of 1955 led to the partitioning of the country, along the 17th parallel, into the Communist-controlled North, and the South under President Diêm. In 1959 fighting broke out between Communist insurgents in the South and the Diêm government. Kennedy, in support of the South, agreed to send military "advisors": Initially a few hundred, by 1962 their numbers had risen to twelve thousand.

In August 1964 it was alleged that North Vietnamese naval vessels had attacked two US Navy ships in the Gulf of Tonkin. While it is now acknowledged that the attack was misrepresented by the US government, the event provided a pretext for Congress to pass a resolution authorizing the President "to take all necessary steps, including the use of armed force to assist any member State of the Southeast Asia Collective Defense Treaty requesting assistance." This gave the President authority to act without a formal declaration of war. Johnson continued to escalate American military involvement in Vietnam. By 1965 more than 300,000 combat troops had been committed for action in Vietnam.

The war in Vietnam was the first to be broadcast to Americans on nightly news programs; objections to the draft and to the conflict's casualties spawned a vocal anti-war movement that would be bolstered by Martin Luther King's speech, "Beyond Vietnam, A Time to Break Silence," in April 1967.

There were changes in the Soviet Union, too. In October 1964, the Politburo dismissed Nikita Khrushchev, accusing him of everything from rudeness, incompetence and megalomania to growing old. He was forced into retirement, succeeded by Leonid Brezhnev. As Gaddis points out, Khrushchev's fall was symptomatic of a general failure of power in the US and the USSR. Not only were the non-aligned countries becoming more astute at exploiting their positions, but countries formerly dependent on the superpowers were playing to their own weaknesses to ensure support: tails wagging dogs, as Gaddis puts it. Thus the exhibition USIA planned for Montreal was held against a backdrop of increasing anger at the continuation and conduct of the Vietnam War.

The main political fireworks of Expo '67 had nothing to do with the USA, however. The French President, General Charles de Gaulle, timing his official visit to coincide with the Expo, made a speech in Montreal that ended with the words *"Vive Montréal! Vive le Québec!"* (Long live Montreal! Long live Quebec!) He then added, *"Vive le Québec libre! Vive le Canada français! Et vive la France!"* ("Long live free Quebec! Long live French Canada! And long live France!") Although the speech was made spontaneously to the crowds who met him in Montreal, there is clear evidence that he intended from the start to make a stand for the Quebec separatist movement and thereby assert French interests. The result was diplomatic and media uproar, both in Canada itself and the French press. While it is clear that the speech encouraged the French Canadians, it is hard to see what else the General achieved apart from emphasizing his reputation for being as awkward as possible when necessary.

(at left) Buckminster Fuller, Jack Masey, Terry Rankine, Peter Floyd with a model of the dome behind them.

The dome under construction.

> Model for the dome.

PLANNING AND DESIGN

When, in July of 1964, the United States Government officially accepted an invitation from the Canadian Government to participate in Montreal's Expo '67, it would be the second time in nine years that the country would be involved in a "first-category world class" exposition. In 1958, the Department of State was responsible for planning the US Pavilion at the Brussels World Exposition. For the 1967 Expo in Montreal, responsibility for creating a US Pavilion would instead fall to the United States Information Agency, with Jack Masey named Chief of Design.

Within eight weeks of accepting the invitation, a "selection committee" staffed by Masey and John Slocum was already interviewing prospective American architects and exhibition designers to select a "design team" for the US presence in Montreal. Among

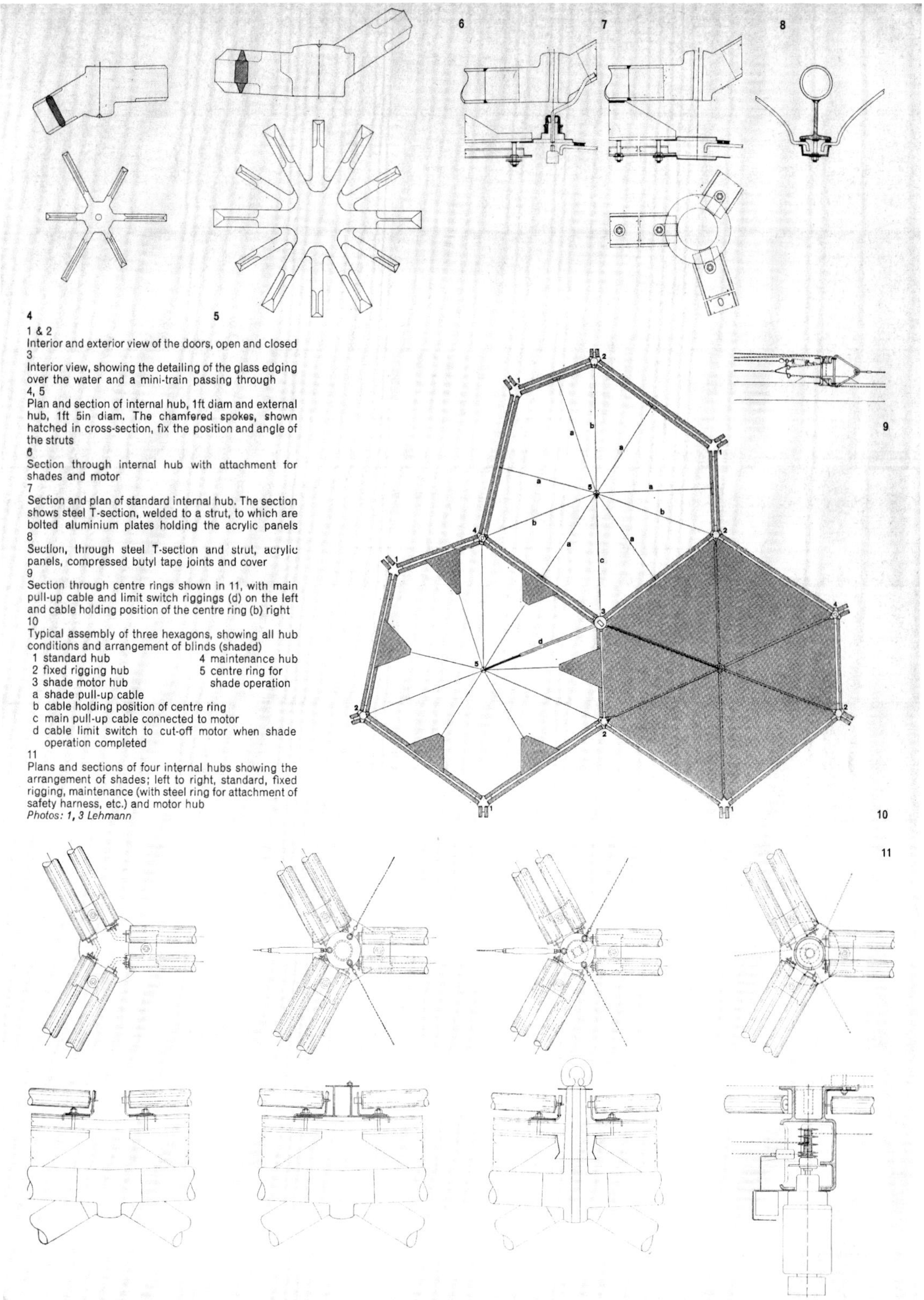

1 & 2
Interior and exterior view of the doors, open and closed
3
Interior view, showing the detailing of the glass edging over the water and a mini-train passing through
4, 5
Plan and section of internal hub, 1ft diam and external hub, 1ft 5in diam. The chamfered spokes, shown hatched in cross-section, fix the position and angle of the struts
6
Section through internal hub with attachment for shades and motor
7
Section and plan of standard internal hub. The section shows steel T-section, welded to a strut, to which are bolted aluminium plates holding the acrylic panels
8
Section, through steel T-section and strut, acrylic panels, compressed butyl tape joints and cover
9
Section through centre rings shown in 11, with main pull-up cable and limit switch riggings (d) on the left and cable holding position of the centre ring (b) right
10
Typical assembly of three hexagons, showing all hub conditions and arrangement of blinds (shaded)
1 standard hub
2 fixed rigging hub
3 shade motor hub
4 maintenance hub
5 centre ring for shade operation
a shade pull-up cable
b cable holding position of centre ring
c main pull-up cable connected to motor
d cable limit switch to cut-off motor when shade operation completed
11
Plans and sections of four internal hubs showing the arrangement of shades; left to right, standard, fixed rigging, maintenance (with steel ring for attachment of safety harness, etc.) and motor hub
Photos: 1, 3 Lehmann

The dome under construction.

Diagram showing how the panels in the dome could be opened or closed to admit light.

the twenty architects and designers contacted were Philip Johnson, Marcel Breuer, Paul Rudolph, Ludwig Mies van der Rohe, Louis Kahn, R. Buckminster Fuller, George Nelson, Donald Deskey and Cambridge Seven Associates.

On December 8, 1964, based on a recommendation by Masey and Slocum, USIA named Fuller, along with Shoji Sadao, as architects of the US Pavilion structure. Cambridge Seven Associates were designated interior architects and exhibition designers. A few months later, USIA asked Congress to appropriate $ 12 million to fund the entire US effort in Montreal. USIA, chronically underfunded by Congress, once again had its request cut: The appropriation for Expo '67 was slashed to $ 9.3 million—nearly 5 million less than the $ 14 million appropriated for the State Department's efforts in Brussels nearly a decade earlier.

Undaunted by the reduced budget, representatives of Fuller and Sadao, Cambridge Seven and USIA met in Cambridge, Massachusetts for a design development meeting in midsummer of 1965. At that meeting, Peter Chermayeff of Cambridge Seven Associates suggested that Fuller's "hemispherical" dome be redesigned as a "3/4 sphere" for Expo '67, a change that would substantially increase the height and drama of Fuller's enveloping geodesic structure. The concept of the "3/4 dome" was quickly accepted by all present.

Once the new shape of the dome was agreed to, Cambridge Seven proceeded to design a nine-story high vertical complex of multi-level platforms upon which exhibits would be displayed. They devised a circulation plan in which visitors, upon entering the US Pavilion, would begin their experiential journey by ascending a 125-foot long escalator to the top platform. There, by means of a series of short escalators, stairways and bridges, visitors would descend through all the individual exhibits. This vertical composition would offer a number of benefits: During the day, it would allow thousands of visitors an hour to move through the exhibits on the various levels of the dome while enjoying spectacular views of the city of Montreal. At night, the transparent dome would be transformed into a "shimmering image" full of movement and color that would become one of the most dramatic and memorable images at the Expo.

The Skytrain monorail, taking visitors around the Expo, passed through the US Pavilion.

> The dome has a golden hue in the evening light.

THE CONTENTS OF THE DOME

The contents of the US Pavilion at Expo '67 were not determined by "committees of outside experts" as they had been for the US Pavilion at Expo '58 in Brussels. Instead, the planning process was confined to USIA itself. In January 1965, to stimulate discussion and solicit ideas, Masey circulated a list among senior USIA officials of proposed exhibit topics that included lunar exploration, American heritage, architecture, film, graphic design, industrial design and contemporary art.

The responses of these officials were varied and curious: one cautioned that the American exhibit in Montreal should not pander to what was purported to be a "sense of low self-esteem" on the part of Canadians *vis-à-vis* the US. Another suggested that there should be a major exhibit on the Tennessee Valley Authority. And one USIA officer wrote: "We all know Canada is a long, thin country and that the people of British Columbia may have a lot more in common with the people of the State of Washington than they do with their fellow citizens in Quebec and would be happier if there were a little more corn, dear to the hearts of Canadians. I would like to see Texas Longhorn or Santa Gertrudis cattle. I would like to see a slice of a giant redwood."

Determining what the contents should be was made even more difficult since it was expected that sixty per cent of the visitors to the Expo might be Americans, and inevitably some would think of its contents that wrong choices had been made. (Indeed, so it proved. While there were but a handful of anti-war protesters, there were complaints from visitors and the American press about the exhibits.) With the Expo taking place on America's doorstep—it would take two hours to fly to Montreal from New York, four from Washington or Chicago—USIA planners were faced with tough questions: At which audience should the US Pavilion be aimed—the Canadian? The American? An international audience?

By mid-1965, content planning was moving along swiftly with Masey in almost daily contact with Ivan Chermayeff, Cambridge Seven's key exhibition designer. Also in the loop was Terry Rankine, an architect with Cambridge Seven, whose role was to "marry" whatever content ideas might emerge to the vertical multi-platform system that formed the key circulation hub in the dome.

According to Ivan Chermayeff: "Three of us, Jack Masey, Terry Rankine and I constituted the main [creative] team ... The first problem with which we concerned ourselves was what we wanted to say and to whom. It seemed correct to show the United States to foreign visitors and in a light unfamiliar to them ... We gathered

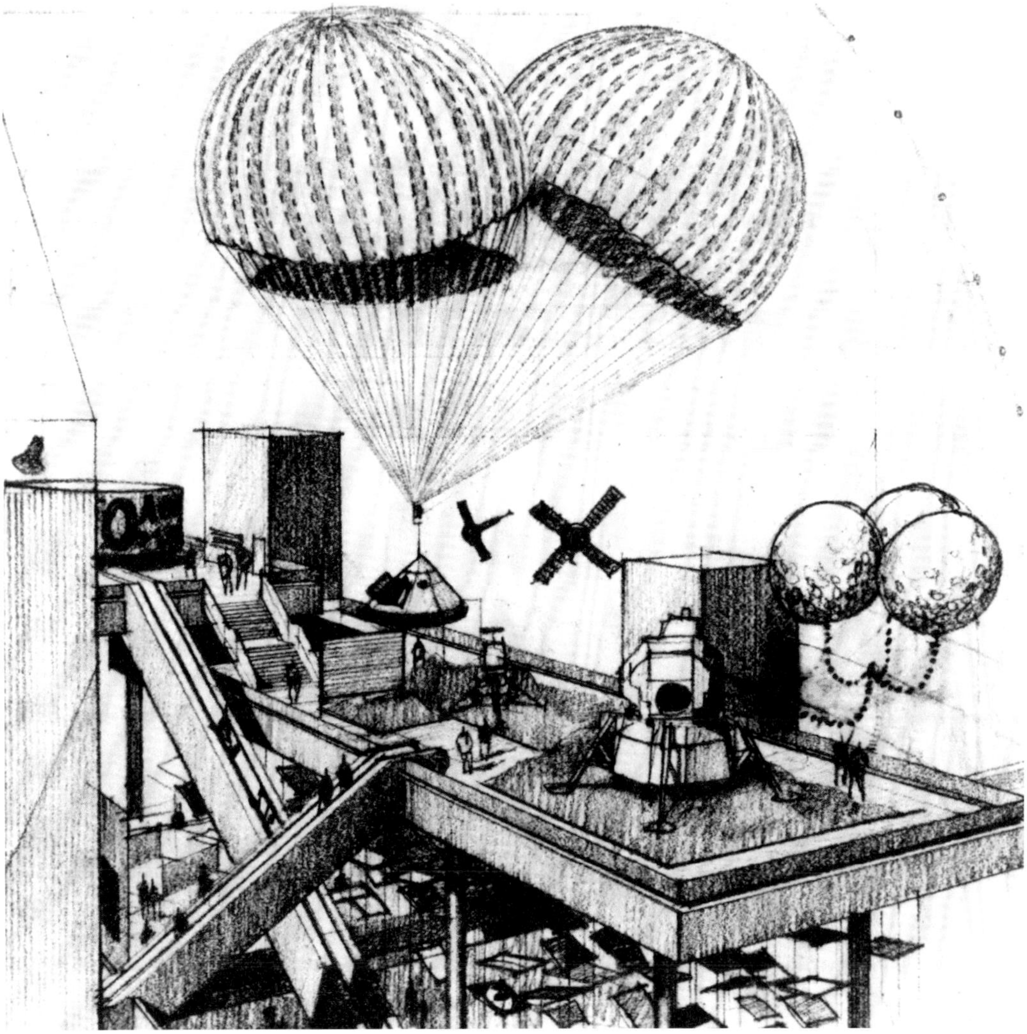

Design drawings emphasized
the role of space exhibits.

together a series of varied exhibits subjects connected only by virtue of their touching on aspects of the American creative spirit ... The force of these exhibits ... was more visual than rational. We chose subjects for exhibits for love, for their symbolic value and for their beauty and craftsmanship ... we excluded electric ice-crushers, kitchens of tomorrow or the products of Detroit. The exhibits we chose gave us an opportunity to show aspects of America related to people's emotions, memories and experiences which would enlighten people about themselves. We hoped ... they might be moved to see things as they had never quite seen them before."

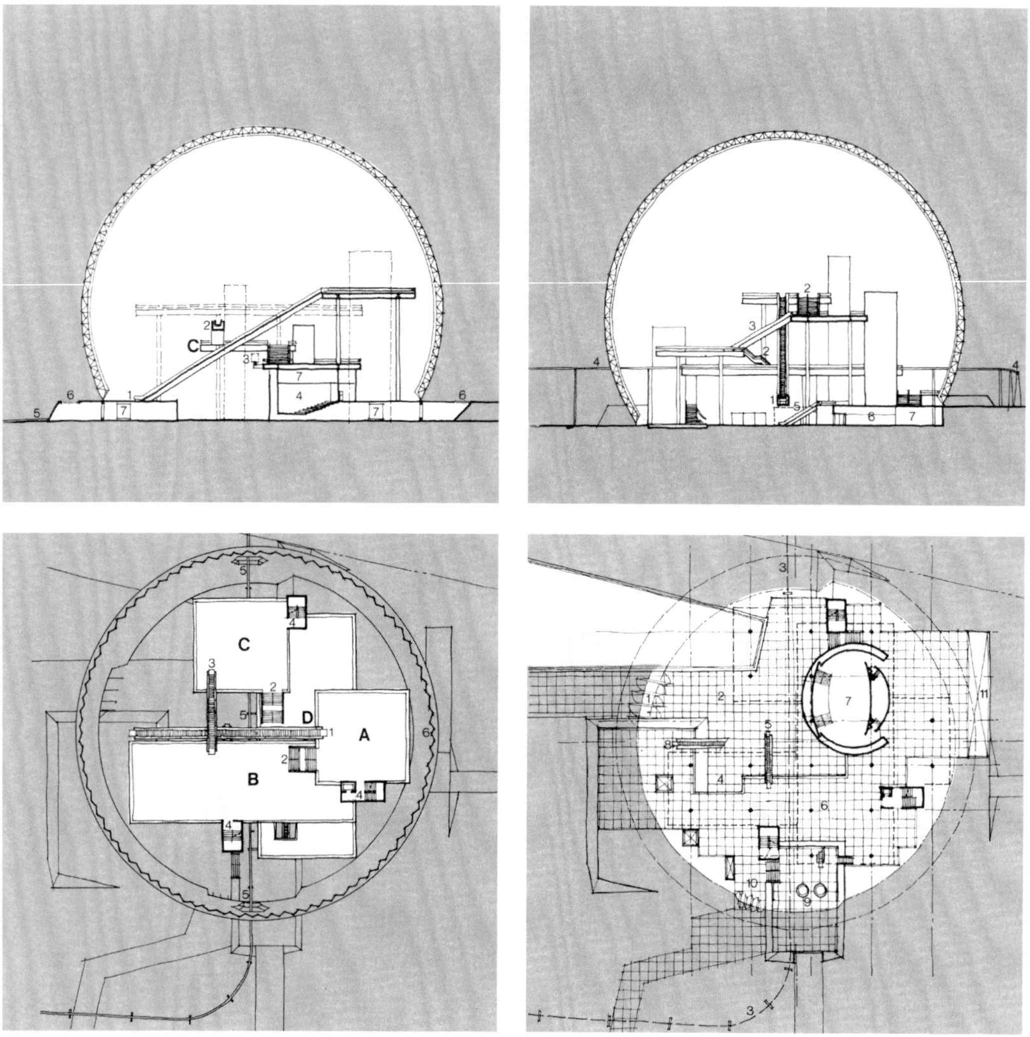

Sectional drawings and plans showing the locations of different exhibits.

"We tried to create the atmosphere of the times in America that people could recognize quickly. And so we had parachutes and space capsules, Hollywood films and Pop Art and the kinds of things that we thought would represent an atmosphere of life in America at the time."

Ivan Chermayeff, Designer

A model of the exhibition area.

> Looking down into the exhibition from the top level where visits started.

THE DOORS OPEN

When the Montreal Expo opened to the public on April 27, 1967, visitors entering the US Pavilion found themselves in a twenty-story high, steel and acrylic "geodesic sky-break bubble." Greeting them in the glistening transparent dome was a Piranesi-like panorama of criss-crossed escalators moving people up and down and punctuated by images of Native American headdresses, fifty-foot high Pop Art paintings, umbrella-like parachutes overhead supporting space capsules, and giant photos of Marilyn Monroe and Marlon Brando.

The 125-foot long escalator—at the time, one of the longest free-span escalators in the world—transported a steady steam of visitors to the topmost platform, some nine stories above the main entry level. Here, visitors could not only view a panorama of the city of Montreal, but could also see the varied contents of the dome on its multi-leveled platforms below.

The first major exhibit on the top platform was "Destination Moon." Devoted to telling the story of the "Apollo" Program (developed as part of President Kennedy's goal to land a man on the Moon by the end of the 1960s), the exhibit featured some of the actual vehicles the US had flown into space. Suspended from the

The escalator bringing visitors to the top level to start their visit was the longest constructed at the time.

roof of the dome by open parachutes were an Apollo Command Module, Alan Shepard's Mercury Capsule and Gordon Cooper's Gemini Capsule. Also suspended high above visitors' heads were the gleaming silver weather balloons of the Explorer Program. As visitors began their descent, they encountered a simulated lunar landscape, as well as exhibits of spacesuits and training couches.

Continuing downward, the next major exhibit was "American Painting Now" devoted to the Pop Art movement that had begun in the United States in the 1960s. Here, visitors saw twenty huge paintings (ranging in height from ten feet to five stories) commissioned for USIA by Alan Solomon, former Director of the Jewish Museum in New York. On display were works by leading contemporary painters and Pop artists including Claes Oldenburg, Jasper Johns, Jim Dine, Andy Warhol and Robert Rauschenberg.

Parachutes for landing a re-entering space vehicle formed a central visual motif.

UNITED
STATES

The space exploration exhibit featured a real lunar module at a visualization of Tranquility Base.

Descending once again, visitors reached "The American Cinema" exhibit where they could view sequences from Hollywood films going back to the silent film era, amidst giant photographs of Hollywood film icons including Mary Pickford, Marlene Dietrich, Clark Gable, Elizabeth Taylor, and others. Among the most popular of the actual movie props displayed was an engineless yellow cab that had been seen in hundreds of Hollywood films. The exhibit also included the chariot from the 1926 movie *Ben Hur* and the catapult used to storm the Bastille in the 1935 movie *A Tale of Two Cities.*

The final major exhibit visitors encountered on their downward journey from the top of the dome was "The American Spirit" display, designed to highlight the ingenuity and creativity of Americans throughout their emergence as a nation. On display were a wide variety of handcrafted objects: Navajo silver bracelets and wooden "Santos" from the Southwest, totem poles from the Northwest, quilts from Pennsylvania, cowboy parade saddles with gold and silver overlay, and guitars played by Elvis Presley and Burl Ives. Hand-sewn Raggedy Ann dolls, weather vanes, decoys, early mousetraps, and early campaign torches were included to represent the inventive spirit of previous generations of Americans.

Giant images of Hollywood stars dominated the cinema display.

"In form and philosophy, the building and its display say 'today' with wit, profundity, pertinence and charm... The result is a superb demonstration of American culture, now, that makes it clear that this is what 20th century culture is about, as well."

The New York Times

The New York yellow cab was used in many Hollywood films.

> The scale of the cinema display.

>> The large-scale artworks were commissioned for the Pavilion.

"Please change the U.S. Pavilion at Expo '67. It is utterly terrible. Is the U.S. known for ... Raggedy Ann dolls and Hollywood? We are known for our great power and scientific knowledge."

Letter to *New York Daily News*

Displays of everyday objects disappointed some visitors.

1840

"A TIME TO PLAY"

USIA commissioned American photographer Art Kane (who had been in Masey's wartime camouflage unit) to produce a film that would be in keeping with the US Pavilion's theme, "Creative America." Kane, a still photographer who had never previously made a film, chose universally-recognized children's games as a theme that would reflect the competitive nature of adult society. Like so many of the exhibits, the film was intended to be "easily understandable by emotion and the senses alone," to quote Ivan Chermayeff. Kane filmed American children of many races playing games that reflected the joys, sorrows, insecurities, warmth and fellowship of adult life. A review of his unconventional film—shown on three-screens—observed that Kane's movie was "neither a lesson nor a form of entertainment, but a sensitive emotional experience."

Art Kane's specially commissioned film explored the world of play.

UNIFORMS AND GUIDES

By mid-1966, there were already indications that the size of the crowds that might visit the US Pavilion at Expo '67 could number as many as five thousand per hour. Anticipating that crowd control would be a major concern when the Pavilion opened, USIA planned to station bilingual guides at key points throughout the exhibits to direct traffic. It was clear that the guides would need uniforms that would make them visible and readily identifiable, but the reduced funding given USIA by Congress had left no money in the budget for them.

And so Masey called American designer Bill Blass, who, like Art Kane, had also served in his wartime camouflage outfit, asking Blass to design uniforms for the US guides "for free." Blass readily agreed and immediately went to work. His designs—red, white and blue dresses for the women complemented by blue blazers for the men—were among the most stylish at the Expo. Bill Blass was also instrumental in persuading Burlington Industries, a large apparel manufacturer, to provide the uniforms at no cost.

The uniformed guides were bilingual in French and English.

A handful of Vietnam War protestors were among the visitors.

REACTION TIME

Shortly after Expo '67 opened to the public, reaction to the US Pavilion was mixed: Initially, many American visitors found fault with what was shown in the US Pavilion—as had been anticipated, and some put their complaints in writing. One, with no return address, told Masey where he could put the Raggedy Ann dolls, another complained, "Is the US known for Indians, cowboys, Raggedy Ann dolls and Hollywood? We are known for our great power and scientific knowledge."

The American press frequently reflected these sentiments: The *Birmingham Sun-Bulletin* said, "The exhibit inside the dome is not frivolous, but

Another highlight of Expo '67 was Moshe Safdie's acclaimed Habitat housing.

scandalous." The *Washington Star* observed: "The net effect of the US pavilion is one of gawky self-consciousness … the disproportionate emphasis … on aging film sirens … can only tend to reaffirm the shopworn cliché once cherished by all foreigners—that American culture is composed of movies and chewing gum." The *Charleston Gazette* asked: "What the hell does all this mean?"

Canadian visitors, on the other hand, reacted very favorably to the contents of the US Pavilion as did the Canadian and international press. The *Montreal Star* said: "The fact is that the American Pavilion … is possibly the most brilliant manifestation of man and his world at Expo '67." *Frankfurter Allgemeine Zeitung* said: "Wit, elegance and irony are best accomplished in the American exhibition which we look upon as the sensation of Expo '67. There is no boasting about technical achievements, nor about industrial products; the largest industrial nation in the world does not exhibit one single automobile … they are not trying to educate, to boast; they are just pleasing. [The US exhibit is] a masterpiece of intelligent wit and most pleasing self-irony."

Reaction to the US Pavilion at Expo '67 within USIA itself was also mixed: The official USIA report at the conclusion of Expo '67 evaluated the effectiveness of the exhibition: "It appears to be the consensus—at least among [USIA's] career Foreign Service Officers assigned [to the US Pavilion]—that, apart from the exhibit concept itself, little attention was given to USIA objectives and that political consciousness of the opportunity provided by an International Exhibition was not evident. The feeling is that the exhibition should have been used deliberately and consciously as an instrument of US foreign policy."

The same report also notes, however, that "USIA … found itself in the unprecedented situation of having to contend with Americans on foreign soil as a major public opinion factor. This exposed … [the USIA staff] … to the nerve-wracking novelty of cultivating the positive disposition of non-American media while fending off as graciously as possible the puzzled or hostile sallies of the American press and visitor alike."

For many Americans, the trip to the Canadian Expo would be their first trip outside of the United States. It would also be their first encounter with a US Government exhibition that had been designed for a non-American audience. (USIA was prohibited by law from attempting to influence American audiences. The limitations imposed on USIA by the Smith-Mundt Act were so restrictive that, in 1967, most Americans had never even heard of USIA.)

Many American visitors feared that the US Pavilion at Expo '67 would be subject to ridicule by foreign audiences, particularly when compared to the massive Soviet presence there. The *Lowell Sun* summed up their concerns: "While the United States Pavilion ... [has] little to show of what the country stands for in today's world ... Russia features displays showing how the USSR is harnessing nuclear energy for peaceful purposes and achievements in industry, agriculture, medicine, science and education."

As word of the overwhelmingly positive reports in the foreign press of the US Pavilion at Expo '67 filtered back to the US, however, these fears subsided significantly. Nine years after the closing of Expo '67, on the occasion of the celebration of the American Bicentennial in 1976, the *American Institute of Architects Journal* published a list of "the proudest achievements of American architecture" in the country's two-hundred-year history. Among the buildings cited was the US Pavilion at Montreal's Expo '67. On the same AIA journal list would be the USIA's next major exhibition project: the Unted States Pavilion at Expo '70 in Osaka, Japan.

The grandiose Soviet pavilion.

Photographic imagery and a portrait of Lenin in the Soviet pavilion.

> The hammer and sickle outside the Soviet pavilion contrasts with Bucky's bubble.

1967
ВСЕ
ИМЯ
1917
OF MAN
OF MAN
TOUT POUR
L'HOMME
POUR LE BIEN DE
L'HOMME

KIMONOS AND MOON ROCK

UNITED STATES PAVILION
JAPAN WORLD EXPOSITION
OSAKA, JAPAN
1970

CARPENTER

In the 1970 iconoclastic film *M.A.S.H.*, two of the doctors, Hawkeye and Pierce, get the opportunity to visit Japan to treat the injured son of a Congressman. Though set during the Korean War, the film was immediately recognised as a black farce about Vietnam, and, in both conflicts, Japan was a staging post for troops and equipment in transit, and a coveted R&R destination. Japan's role in the Cold War was not just transient, however. The economic recovery of the country after the Second World War was, by the end of the 1960s, turning from a regional success story to a global phenomenon. In electronics, Sony had pioneered the transistor radio fifteen years earlier and continued to innovate; Honda motorcycles and Toyota cars were penetrating international markets. America had looked to Japan to provide economic stability in the region to counter the potential power of China: this had largely been achieved. Participation by the USA at a World Exposition in Osaka was a political necessity for such an important ally and trading partner.

America, in particular, needed to make the case for its policies in the Far East, where the escalation of the Vietnam War was a major concern. Elsewhere in the world the bubble of unrest that had burst in the May 1968 riots in France, the American Embassy demonstrations in London, and student protests in Berlin, had been seen by a worldwide audience. The savage repression of the Prague Spring in the same year had been widely condemned. Protests continued against the Vietnam War in 1969 and 1970. In the United States, student protests had turned ugly with the killing of demonstrators by the National Guard at Kent State University, and opposition to the war continued to be linked with demands for civil rights.

The central theme of the Japan World Exposition in Osaka was "Progress and Harmony for Mankind." This theme was projected through four sub-themes: "Towards fuller enjoyment of life, towards more bountiful fruits from nature, towards fuller engineering of our living enjoyment and towards better understanding of each other." These grand aims can also be read as a reflection of the issues facing Japan itself, a list of the challenges faced by a relatively small, heavily populated island country with a strong technology base, looking towards the future but with a recent history of war with its neighbors. Hosting a World Exposition was an opportunity to show the new Japan to the world. For the USA, participation was an opportunity not so much, as in the past, to confront

< A young visitor sits in Virgil "Gus" Grissom's training couch.

Communism but to present a positive image of America to the region. On film, *M.A.S.H.* sought to show, beneath a sparkling surface of provocative hippy comedy, how a group of disparate Americans confronted and learned from their often involuntary involvement in war: one could draw an analogy between the movie and the intention behind the US Pavilion to present a broader picture of the reality and variety of life in America beyond the news footage of Vietnam. As it turned out, the hippies came along as well.

Planning for participation at Osaka began in 1967: USIA planners suggested a selection process to pick architects and designers in July, and proposed that an "exhibition design team," consisting of architects and designers from outside the US government, be created. The purpose in creating the design team concept was to bring together, at the very outset of the project, the architects of the Pavilion enclosure with the designers of what was being enclosed, to insure that optimum design homogeneity prevailed.

It was proposed to organize an invitational competition to select the final design team, which would consist of a combination of architects and exhibition designers. Six architects were selected, and invited to make a proposal in partnership with an exhibition designer: six exhibition designers in turn were invited to propose, with chosen partner architects.

Among those participating in the various design teams who made presentations were: Philip Johnson, Paul Rudolph, James Stewart Polshek, George Nelson, Isamu Noguchi, Ivan Chermayeff and Thomas Geismar. In the end, eleven teams submitted proposals to an advisory panel consisting of private citizens and government representatives in the fields of communications and the arts. They were invited to review the submissions and select an exhibition design team. The USIA was represented by Howard Chernoff, Commissioner General for the Pavilion, Robert Sivard, Art Director, and Jack Masey, who had been appointed Deputy Commissioner General of Design for the event. The panel recommended the team of Davis, Brody, Chermayeff, Geismar and de Harak to design the US Pavilion in Osaka. Their winning proposal was for a balloon-like, air-supported structure, 275 feet in diameter, initially intended to function, in part, as a giant projection screen. Given its height and shape it was nicknamed "the Pumpkin."

A proposed budget of $ 17,750 million was presented to the Bureau of the Budget in February 1968; just under sixteen million dollars was approved for presentation to Congress. Planning continued on the basis of this figure, but in June, Congress cut the funding to ten million dollars, a decision the Senate refused to overturn. A year's worth of planning and development was effectively lost. This regular practice by different factions in Congress of slashing the USIA budget, both overall and on specific projects, is perhaps a final reflection of isolationism. It also represents the isolation of the USIA; as a government agency forbidden to operate within the USA itself, it had no homeland constituency to come to its defense.

While the designs were being proposed and developed, the budget was successively cut. As in all previous USIA exhibitions, the initial budget request to Congress for sixteen million dollars was cut and cut again, to a final appropriation of ten million dollars. Each time the budget was cut, designs for the pavilion and exhibits changed because the administrative costs could not be reduced. As the budget shrank, so did the height of the building. The initial 30 story high "Pumpkin" was redesigned as four smaller balloon-shaped structures. After the last budget cut, it was deflated once again: the final design became the two-story high "Band-Aid".

The intent of the designers was to create an air-supported, easily constructed structure that could accomodate enormous crowds. The initial "Pumpkin" shape created a form that provided the combined benefits of a circle and a square. The subsequent "four-ball" scheme, as it became known, was meant to have films shown continuously inside each ball. The final design, created after the budget was reduced to ten million, was partially subterranean: its low inflated roof was not a suitable surface for projected images, and its lower level was to be created out of bermed earthworks.

The use of this innovative inflatable technology (which was being explored at the time by other architects such as the Archigram Group in London, though not on such a scale) was one of the factors in the panel's decision. The Pavilion as built was the largest clearspan air-supported roof in the world. The structural engineer who conceived the successive schemes was David Geiger (a partner in the New York firm of David Geiger and

Horst Berger) who acted as a consultant to the design team. Geiger was recommended to the team by Mario Salvadori, a pioneer of fabric structures who had designed smaller inflatable structures for the 1939 New York World's Fair. As Alan Schwartzman, a member of the design team, has pointed out, it was Geiger's ingenuity that made the design possible. The vinyl-coated fiberglass membrane roof allowed sunlight to pass through into the interior spaces below by day, and its interior light made it glow at night.

Seen from a distance on the ground, the Pavilion appears understated: a low, silvered curve. Only when the visitor entered the structure did the size and elegance of the elliptical plan of the roof become apparent. It was 465 feet (141 meters) long by 274 feet (83 meters) wide. Unlike the highly visible appearance of some of the seventy-one other national pavilions at the Fair, the US Pavilion made a quiet affirmation of technological prowess.

BUILDING THE STRUCTURE

It became clear early on that using this new technology on such a scale would require the involvement of a skilled construction firm, and USIA interviewed a number of major Japanese construction companies before selecting Ohbayashi-Gumi of Tokyo as the main contractor. The delays imposed by budget cuts meant, however, that the detailed price negotiations had to take place without anything more than preliminary drawings, and in addition USIA wanted a fixed-price contract to avoid cost overruns. Ohbayashi, aware of the prestige of the project, came up with a novel solution. They would agree to a fixed-price contract but only if their engineers worked alongside the design team to develop the final drawings. This was a practical solution that saw twenty-four Japanese engineers and designers arriving in New York to help develop the final design.

The novelty of the design solution posed other problems. The inflatable structure would have to be extensively tested with an inflatable, scaled mock-up, as well as being rigorously tested on site before the exhibits were installed. The Japanese authorities were concerned about possible injuries or worse resulting from a collapse of the roof, and the final interior structure was, in part, configured to support the roof should it indeed fail.

Model of final design of air-inflatable US Pavilion at Expo '70. 260 foot wide by 470 foot long, it was nicknamed "the Band-Aid".

Davis, Brody, Chermayeff, Geismar and de Harak's original design, nicknamed the Pumpkin, that won the closed competition for the design.

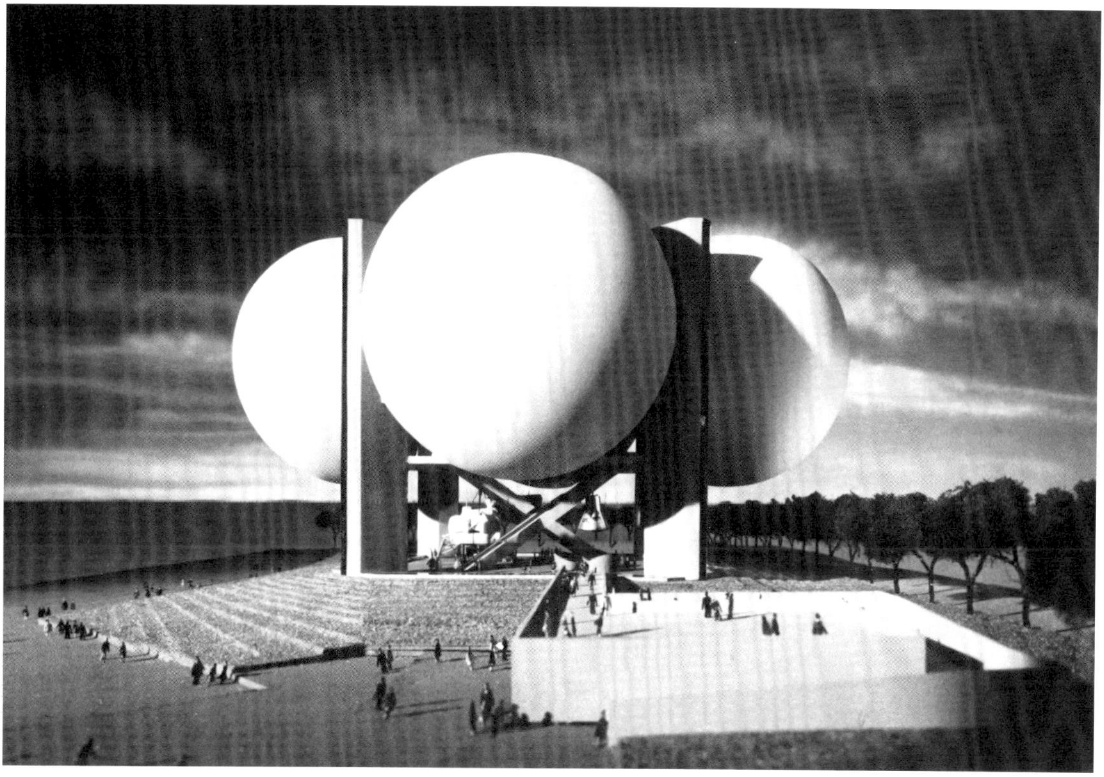

The second stage "four-ball" design.

Design Competition entries:

Myron Goldsmith, Skidmore, Owings & Merrill, architects; Morton Goldscholl Design Assocs., designers

Isamu Noguchi, sculptor; Pavilion Associates, architects and designers

Helmuth, Obata & Kasselbaum, architittects; Ronald Beckman, Research & Design Inst., designers

William J. Mouton, Jr., engineer; Philip Johnson, architect, Harper & George, designers.

Gruzen & Partners

George Nelson & Co., architects and designers

Minoru Yamasaki & Associates, architects

James Stewart Polshek, architects; Arnold Saks & James S. Ward Inc., designers

Paul Rudolph, architects

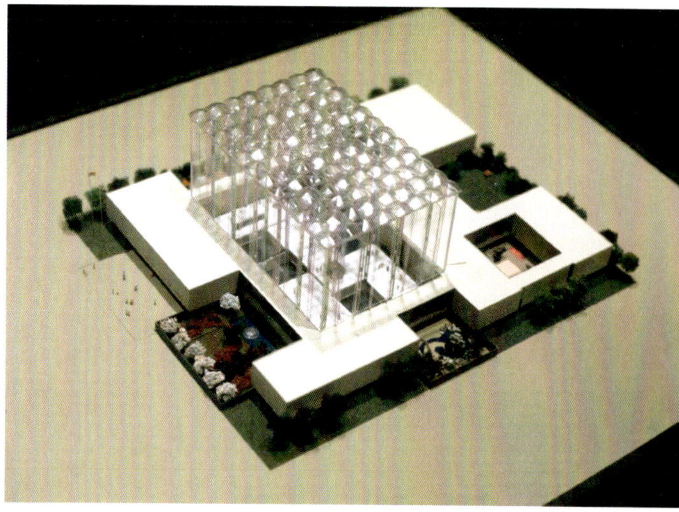
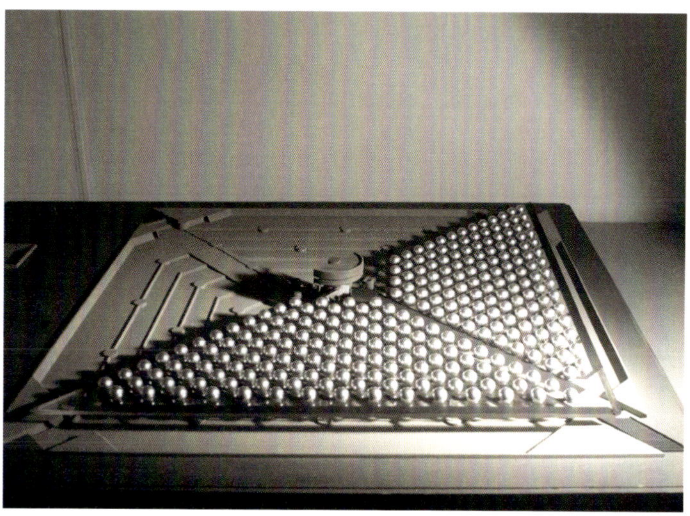

They even sent a man with a samurai sword onto the completed roof to cut a hole in it: the resulting blast of escaping air knocked him off his feet, but the roof remained intact. The roof was described by David Sutton, a member of the design team, as "the world's biggest trampoline," and indeed, jumping onto the roof and bouncing around became something of a relaxation exercise with the design crew—until the authorities told them to stop.

The roof membrane was attached to high-strength steel cables that were in turn secured to a funicular reinforced concrete ring at the base of the berm. These cables criss-crossed the roof structure forming a diamond pattern. They insured the stability of the roof but did not support it as such. The cables created, together with the tension ring, a structure that would resist earthquakes and typhoons.

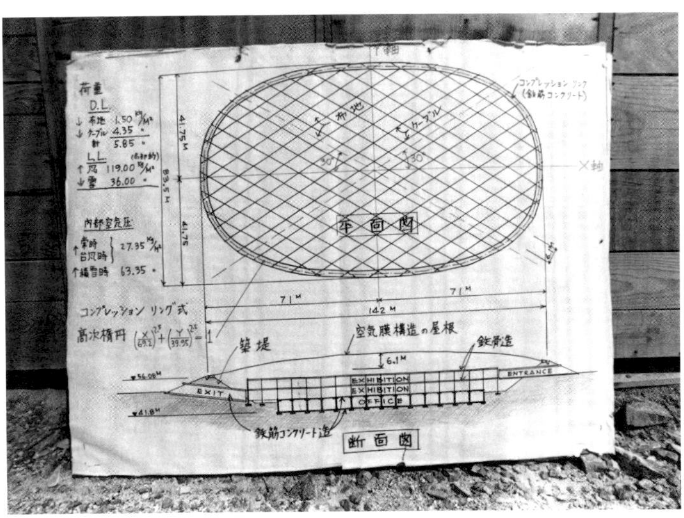

The design team at lunch.

Workers' drawing
at construction site

Model of the roof design.

The scale model for
testing the roof.

Air pressure within the dome was maintained at three-hundredths of a pound per square inch above standard pressure (fifteen pounds per square inch at sea level), using two large blowers, with a further two as a reserve. Air pressure was further maintained by using revolving doors as air locks at the entrances and exits. Such doors were, however, unfamiliar to the majority of Japanese visitors: not only did queues build up as visitors had to be shown through them, there were initially breakages, until the doors were replaced by heavy-duty plate glass. Showing visitors through individually meant a flow rate of about 8500 per hour: this was about the maximum capacity of the pavilion, for which the planned visit time was around an hour.

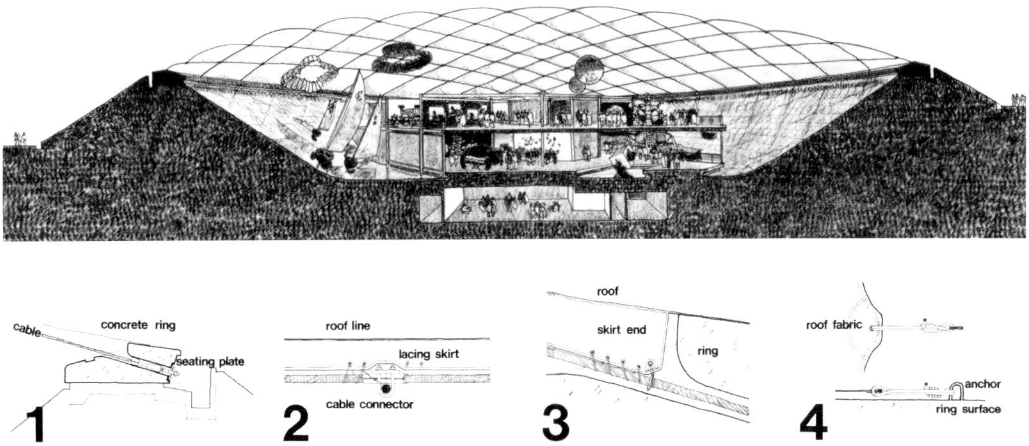

Section drawing of the design and details of the roof cabling.

Sewing the roof fabric.

Roof fabric and cables.

"Whenever a member of the design team came over from New York, their first stop was at the site to launch themselves onto the roof—the world's biggest trampoline!"

David Sutton, Exhibit Designer

Workers laying out the roof.

Positioning the roof.

Installing the steelwork for the exhibits.

The interior being inspected before exhibits were installed.

> Aerial view of the Pavilion after the Fair opened.

>> Visitors queuing to enter the US Pavilion.

DESIGNING THE EXHIBITIONS

Beyond the question of what to exhibit, two main challenges faced the exhibition designers. Firstly, how could the design deal with the enormous crowds that the pavilion was expected to attract. (65 million people were expected to visit the Osaka Expo.)

In order to move the visitors in a safe and steady manner, it was decided to route all traffic in a one-way direction from entrance to exit. This one-way traffic-flow would also structure the visitor experience. Aisles were made deliberately wide so that visitors who wished to enjoy a specific exhibit could linger without delaying others.

The second challenge was more complex: how to present material and information across a language and cultural barrier, since most Japanese spoke no English and were under-informed about American culture and lifestyle. There were, however, some approaches that could be followed. The quintessentially American sport of baseball was very popular in Japan. In addition, craft and folk arts often spoke to each other across cultural boundaries: there was a strong Japanese tradition of decorative art. Photography was a similarly universal medium. There was also continuing global excitement about the space race, and the possibility of a successful Moon landing before the opening of the Fair: the Eagle did land at Tranquility Base in July 1969.

The theme chosen by USIA and the design team for the US Pavilion was "Images of America." Seven main exhibits were chosen: Ten Photographers, American Painting, Sports, Space Exploration, Folk Arts, Architecture, and New Arts. Each exhibit would have its own section of the pavilion, with the space exploration section being visually dominating, as it could be seen both at the lower level and from the upper level platforms.

Principal members of the exhibition design team were responsible for the design of individual exhibits: de Harak was a passionate sportsman and so designed the Sports Exhibit, Geismar was responsible for the Space Exploration Exhibit, Chermayeff for the Architecture Exhibit, and so on.

Views of visitors on the Fairgrounds.

AUTHENTICITY

A principal objective of the design of all of the exhibits was to show real objects, not mock-ups or models: the Apollo capsule had actually flown in space, the baseball bat had actually been wielded by Babe Ruth, and so on. Such authenticity was in the design team's view essential to the success of the whole presentation, which would conclude with the most authentic object of all, a sample of moon rock recovered by the Apollo 12 astronauts during their visit to the Moon. It was to become the most important and exciting object of the whole Fair.

Axonometric drawing of the exhibit structure.

Model of the exhibit structure.

Diagrams showing the location of different exhibits on upper and lower levels of the Pavilion.

> Interior view.

upper exhibition level

scale 0 10 20 30 meters / 20 50 100 feet

legend

a space
b sports
c folk arts
d images of america
e american painting

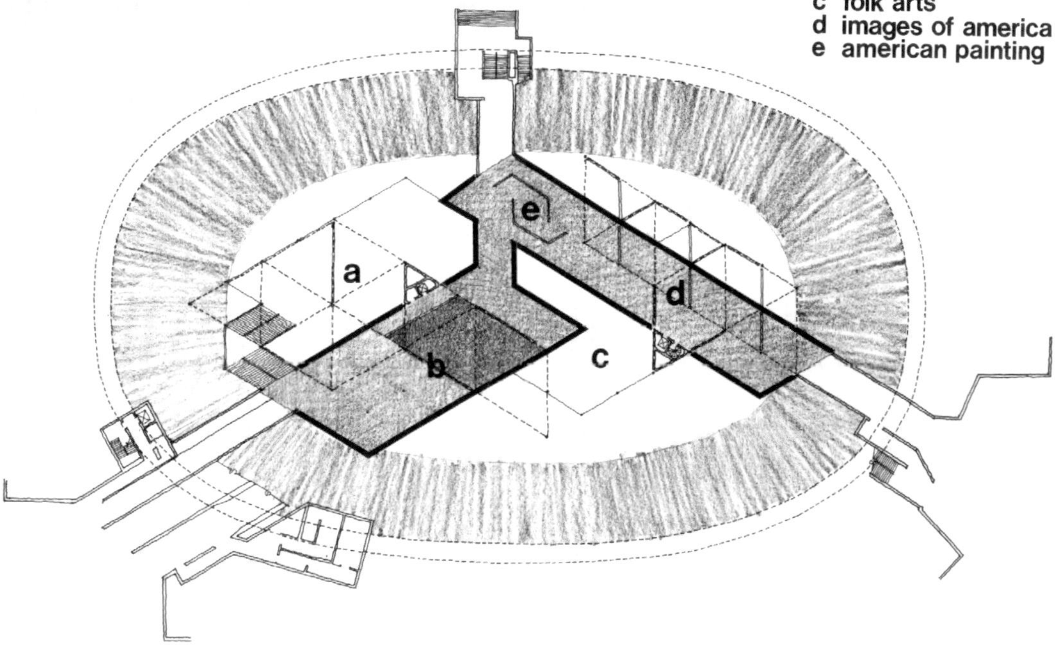

lower exhibition level

scale 0 10 20 30 meters / 20 50 100 feet

legend

a space
b architecture
c folk arts
d new arts

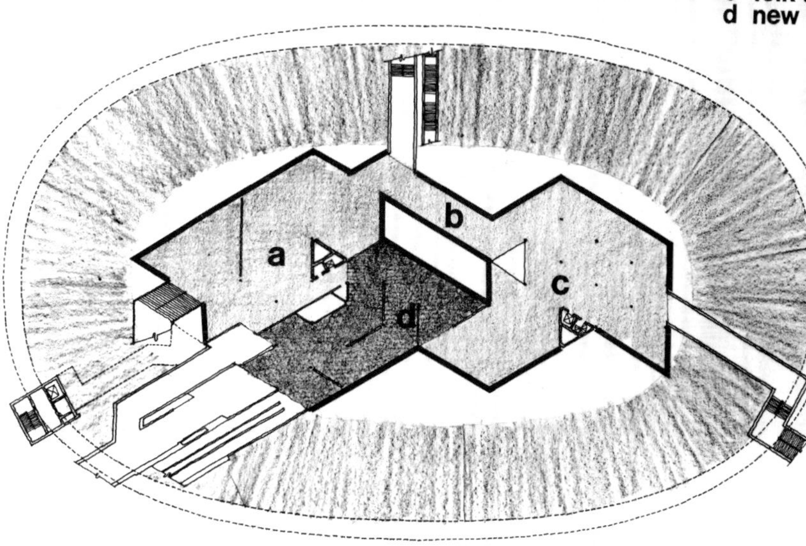

"The Ten Photographers exhibit was not about travel pictures—it was about what it revealed about Americans, it was about the diversity of the American population."

Ivan Chermayeff, Designer

TEN PHOTOGRAPHERS

The photographic exhibition was selected by John Szarkowski, Director of Photography at the Museum of Modern Art in New York. Each photographer was represented by ten images, shown as large backlit black and white transparencies. The selection ranged from the classic landscapes of Ansel Adams through to the then new and important school of street photography represented in the work of Gary Winogrand and Diane Arbus. A balance was struck between urban and landscape photography by including landscape works by Paul Vanderbilt and William Garnett, and works from New York and other cities by André Kertész and Bruce Davidson. Overall, the selection of these documentary photographs gave a varied and deep portrait of American life. Japanese visitors appreciated them for their straightforwardness and honesty, according to a USIA post-Fair report, while the Emperor of Japan declared that they were "art."

AMERICAN PAINTING

The twenty-one original paintings in this section were loaned by the Metropolitan Museum of Art in New York, and comprised figurative works from the nineteenth and twentieth centuries. The artists represented included Frederick Remington, Maurice Prendergast and Marsden Hartley, from the nineteenth century, and Edward Hopper and Andrew Wyeth among contemporaries. Because of the high value of the pictures, they were mounted in glass cases, carefully lit so as to minimize reflection and glare.

The Diane Arbus photograph that was withdrawn from the Ten Photographers exhibit.

The Ten Photographers exhibit as installed, with images by Arbus, Vanderbilt and Winogrand.

> Visitors looking at the Bruce Davidson's photos of Harlem.

ポール バンダービルト
Paul Vanderbilt

SPORTS

The sports exhibit ranked second in popularity after the space exhibit. While baseball took pride of place (especially with the visitors) there were a range of different objects shown: sports cars (modern and historic), dune buggies, a Harley-Davidson, a gyrocopter, various sailing boats including a catamaran and a Boston Whaler, and on a smaller scale, fishing tackle, golf clubs and football equipment.

The most popular item in the sports exhibit was Babe Ruth's locker in which his uniform and baseball bat were on display. It was akin to a shrine to Japan's baseball fans. For the baseball exhibit, authenticity was the key: also included were items belonging to Joe DiMaggio and Walter Johnson, together with gear from other greats such as Ty Cobb and Mel Ott. Most of the objects were on loan from the Baseball Hall of Fame in Cooperstown, New York.

Historic bats and jerseys from the Baseball Museum.

Babe Ruth's locker.

A guide discussing
a Harley-Davidson motorcycle.

The 1915 Stutz Bearcat sports car.

Visitors at the Folk Arts exhibit.

Various displays including the Shaker "retiring" room, carved and painted toys, saddles, Native American and Eskimo masks and gravestone rubbings.

FOLK ARTS

The Folk Arts exhibit was a way of crossing the cultural divide between the USA and Japan as well as paying homage to Japan's long tradition of craft. The work shown included Native American and Eskimo artifacts such as weaving, carvings, masks, basketry and totem poles. Also on display was 19th century anonymous Americana, featuring vernacular objects such as inn signs and quilts, gravestone rubbings, as well as a room of Shaker furniture. These works were gathered from museums and private collectors across the USA.

Rural iron and copper weathervanes on display in front of the mylar-surfaced berm.

NEW ARTS

The potential of technology to link up with art was a current theme in the late 1960s art world, with major exhibitions of kinetic and op art in Europe. USIA commissioned the Los Angeles County Museum of Art to assemble a number of works from its "Art and Technology" project, in which artists were given an unprecedented opportunity to explore the new world of advanced technology and industry, through collaborations with industrial corporations.

Probably the most popular result was Claes Oldenburg's giant red motorized Ice Bag, but there were also more abstract displays that used lasers and ionized gas, as well as Tony Smith's maze of tetrahedral and octahedral cardboard units.

Two views of Claes Oldenberg's *Ice Bag* in the New Arts exhibit, selected from the Arts and Technology Program, Los Angeles County Museum.

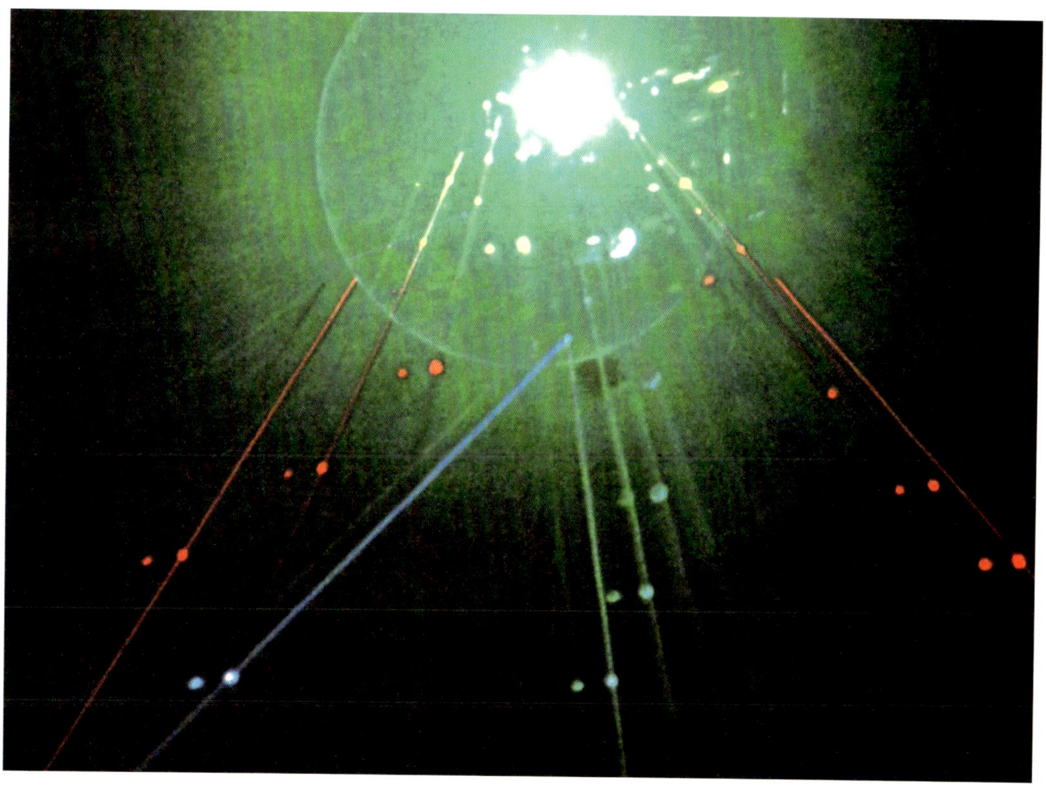

Rockne Krebs' argon and helium neon laser intallation.

Newton Harrison's ionized rare gases in vacuum chambers.

ARCHITECTURE

For the architecture exhibit, curated by Peter Blake, the photographer Elliott Erwitt was commissioned to portray examples of indigenous American architecture, juxtaposed with a wide range of contemporary architecture completed in the previous two years. The large full-color images were presented as backlit transparencies. Among the buildings shown were airports, office buildings, apartment complexes, and museums.

Peter Blake's architecture exhibit featured examples of historic and contemporary American buildings.

SPACE EXPLORATION

The largest single exhibit in the US Pavilion was devoted entirely to space exploration, with emphasis on the Apollo 11 Lunar Mission that had reached the Moon on July 20, 1969. Here authenticity was also the touchstone: NASA agreed to lend genuine objects for display. The spacecraft were either vehicles that had actually flown or were capable of flight. To celebrate the moon landing of the previous year, a major component of the space exhibit was a full-scale simulation of the landing site on the Sea of Tranquility, complete with an actual lunar module and stiffened American flag.

Visitors looking at the reconstruction of Tranquility Base.

Figure of astronaut in an Apollo 11 space suit.

> General view of the Space Exploration exhibit.

Without question, the moon rock on display in the space exhibit (which was retrieved during the Apollo 12 mission) was not only the most popular object displayed in the US Pavilion, but was the most popular object of Osaka's Expo '70.

Nearly four decades later, it is difficult to recall the excitement the Apollo missions generated, and the way in which, for so many people around the world, the lunar landings had a powerful symbolic value. The opportunity to see real space suits and capsules, or to sit—like the Japanese boy in the image at the start of this chapter—in a training seat for a real astronaut, must have been extraordinarily exciting at the time.

Floating figures give a sense of weightlessness.

Tranquility Base.

> Visitors looking at a recovered Apollo capsule.

>> A moon rock captivates a young visitor.

THE SOVIET PAVILION

The footprint of the Soviet Pavilion at Osaka was curved in plan and in elevation. Its arched red spire made it the dominating structure on the Fair grounds. In the eleven years since the exchange of national exhibitions between Moscow and New York, when the Soviet exhibit at Columbus Circle had highlighted the achievements of Sputnik, the Soviet's space program, in contrast to Osaka's thrusting spire, had wilted. The aftermath of problems in Czechoslovakia and elsewhere in the Soviet bloc in the late 1960s continued to plague the Brezhnev era.

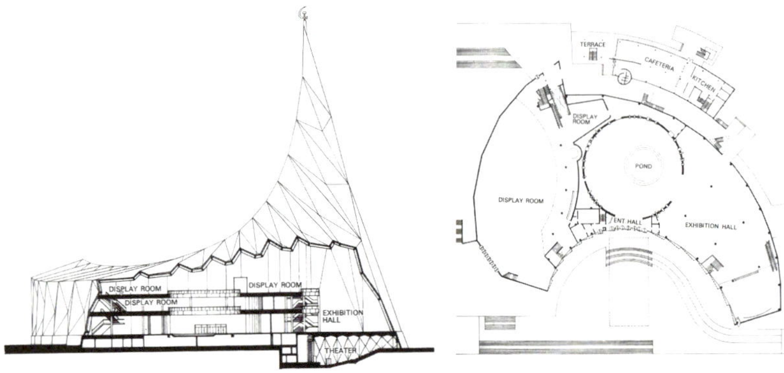

Plan and section of the Soviet Pavilion.

The interior view of the Pavilion with the customary image of Lenin.

Exterior view of the Soviet Pavilion.

IN THE EYE OF THE BEHOLDER

In a post-event report, the USIA expressed the view that participation in the Osaka Exposition had in general been a success, with the space and sports exhibits being particularly well received.

The report concedes that the Ten Photographers exhibit was "the most provocative and controversial exhibit in the pavilion." The work of Winogrand and Arbus, which portrays life on the city streets or people on the margins of society, was challenging at best. But the majority of the complaints were from American visitors, not from the Japanese visitors or press, who, it is claimed, were "overwhelmingly positive."

One image by Arbus, of a young man wearing a straw boater and a button reading "Bomb Hanoi" was withdrawn at the US Commissioner General's insistence, despite a USIA promise to the curator, John Szarkowski, of no censorship. It was taken out because of the sensitivity of USIA to any mention of Vietnam. The goal was to avoid any politicization of the US Pavilion at Expo '70.

There were other complaints, again principally from American visitors, about a group of visiting "hippy" musicians who performed for the crowds waiting for entry to the Pavilion. Since the average waiting time was around two hours, Japanese visitors generally enjoyed the distraction, but there were American visitors who assumed the hippies were part of the official US presentation, and protested accordingly.

Overall, the visual understatement of the US Pavilion was matched by its technical complexity and depth. As Peter Blake wrote in 1970, "the US Pavilion ... is the most daring structure at Osaka: and it certainly impresses the technologists—including those from the USSR." As "Bucky's Bubble" had done in Montreal, the "Band-Aid" at Osaka changed the rules of the game in America's favor.

The moon rock was the most popular exhibit of Expo '70.

AFTER THE FAIR

OPPORTUNITIES LOST AND FOUND 1981 TO PRESENT

The legendary director of the United States Information Agency, Edward R. Murrow, sagely observed: "The really crucial link in the international communications chain is the last three feet, which is best bridged by personal contact—one person talking to another." The USIA exhibits that grew in size and complexity through the Cold War era were spawned by the belief that personal contact—with enemies as well as with friends—was an important element in creating more favorable conditions for stability and peace.

Giant world's fairs were frequently the venues for USIA exhibitions. These world's fairs offered their publics a structured and planned view of the culture, values and aspirations of each exhibiting nation: though the fairs were temporary, they were also fixed in time, simultaneously snapshots and stage sets alike. Exhibitions, often combining real objects with simulated realities—created by use of photographs, graphics, film and full-scale scenic reconstructions—juxtaposed too much imagery in not enough space to create a sense of hyper-reality.

This sense of ideated reality was the product of careful design driven by the need to educate, inform and influence. Because audiences were counted in hundreds of thousands or even millions, the success of any nation's effort at these giant world's fairs is evidently immeasurable, as those who planned, organized and built the exhibitions in this era understood only too well. When President Eisenhower said he expected the 1959 American National Exhibition in Moscow to have a significant effect on the Soviets, what meaning did "significant" have? Rather, what meaning could it have? One thing to remember from our internet-addicted, multi-channeled television era is that these were unique events, and that participation meant presence, real presence in a real place, not in the virtual spectacles that surround us today.

Thomas Geismar regards the Cold War itself, and its two opposing superpowers, as an astringent that clarified the planning and design of exhibitions in the 1950s, 1960s and 1970s, on both sides of the Iron Curtain. Assessing the pros and cons of today's new media as a means of connecting with audiences around the globe, Ivan Chermayeff sees a conundrum: "I am with those who believe that world's fairs are past—that the global nature of the world today, with television and the internet and all that, has made them obsolete. That said, there are

< USIA exhibitions allowed Americans to share candid exchanges of views with foreign visitors.

certain things that you have to feel in your hands—real things. Real things make a difference."

THE WORLD'S FAIR PAVILIONS

The World's Fair Pavilions were the most visible, the most spectacular part of the USIA spectrum, but it was a wide spectrum. Other USIA activities—such as briefing local journalists, arranging cultural, scientific and educational visits and exchanges, meeting local dignitaries in the many countries where USIA had posts—may seem more mundane, but they enabled the organization to assemble a mosaic of opinions and attitudes about the United States from all corners of the world. The assembling process itself contributed in important ways to the formulation of American foreign policy and the maintenance of America's reputation abroad.

The US exhibitions discussed in this book were nourished by the information gathered in this manner by USIA. This information from the field, along with the growing experience of USIA staff, explains how participation at events with very tight timeframes and limited budgets—the Jeshyn Fair in Kabul is one example—could be realized. The professionalism of USIA teams was often matched by the generous support of the US private sector, as was the case at the American National Exhibition in Moscow in 1959.

THE BEIJING PROPOSAL

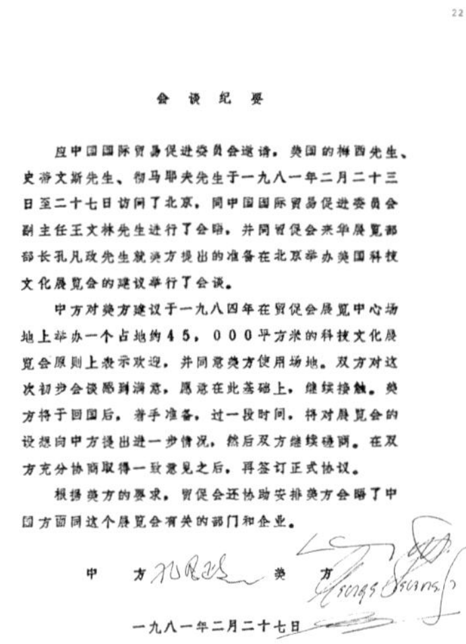

22

会 谈 纪 要

应中国国际贸易促进委员会邀请，美国的梅西先生、史蒂文斯先生、彻马耶夫先生于一九八一年二月二十三日至二十七日访问了北京，同中国国际贸易促进委员会副主任王文林先生进行了会晤，并同贸促会来华展览部部长孔凡政先生就美方提出的准备在北京举办美国科技文化展览会的建议举行了会谈。

中方对美方建议于一九八四年在贸促会展览中心场地上举办一个占地约４５，０００平方米的科技文化展览会原则上表示欢迎，并同意美方使用场地。双方对这次初步会谈感到满意，愿意在此基础上，继续接触。美方将于回国后，着手准备，过一段时间，将对展览会的设想向中方提出进一步情况，然后双方继续磋商。在双方充分协商取得一致意见之后，再签订正式协议。

根据美方的要求，贸促会还协助安排美方会晤了中国方面同这个展览会有关的部门和企业。

中　　方　　　　美　　方

一九八一年二月二十七日

In late 1979, Jack Masey, who had completed his service with USIA and had formed his own design office in New York, met with architect I. M. Pei, film and television producer George Stevens Jr., and designer Ivan Chermayeff to discuss the idea of promoting a possible exchange of national exhibitions between the United States and the People's Republic of China, similar to the exchange of exhibitions between the US and the USSR that took place in 1959.

For the US effort in Beijing, the group proposed an exhibition of science, technology and culture to be shown in the Chinese capital in 1984. The group was invited to China in 1981, but mindful of the difficulties experienced by American negotiators in the Soviet Union in 1959, they expected a complicated reception in Beijing. Instead, they received quite the opposite: an overt offer of cooperation from their Chinese counterparts. In a Memorandum of Understanding that both sides signed three weeks later, the Chinese had identified a site for the American Exhibition in central Beijing. The site proposed by the Chinese measured 450,000 square feet and was a twenty-minute bicycle ride from Tiananmen Square. It would be open to the public for four months; the group estimated that the exhibition could draw as many as 7,200,000 visitors during its 120 day run.

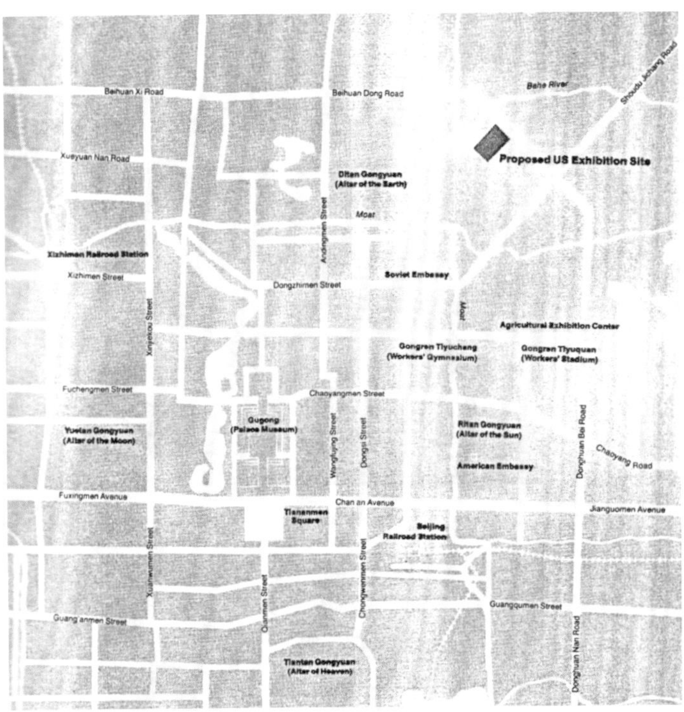

A site near Tiananmen Square was designated for the 1984 US exhibition in Beijing.

TEXT OF CHINA/US MEMORANDUM OF UNDERSTANDING

Following are the English and Chinese versions of the Memorandum of Understanding which was signed in Beijing on February 27, 1981. Signing for the Chinese side was Kong Fanzheng, Director of the Department for Foreign Exhibitions in China of the CCPIT, and for the American side, Ivan Chermayeff, George Stevens, Jr. and Jack Masey of the US Exhibition Design Group.

MEMORANDUM OF UNDERSTANDING

At the invitation of The China Council for the Promotion of International Trade (CCPIT), Messrs. Jack Masey, George Stevens, Jr., and Ivan Chermayeff of the United States of America visited Beijing from February 23 to 27, 1981, and met with Mr. Wang Wenlin, Vice President of CCPIT. They held discussions with Mr. Kong Fanzheng, Director of the Department for Foreign Exhibitions in China of CCPIT, on the project of holding a United States Science, Technology and Culture Exhibition as proposed by the U.S. side.

The Chinese side expressed welcome in principle to the holding in 1984 of the proposed U.S. Science, Technology and Culture Exhibition, which would cover an area of about 45,000 square meters on the site of the CCPIT Exhibition Center, and agreed to the U.S. side using the site. Both sides were satisfied with the preliminary talks held between them and desired to continue contacts between the two sides on this basis. The U.S. side stated that further preparations will be made after their return to the U.S., and additional information will be forwarded to the Chinese side. Following this, further negotiations between both sides will be held. A formal agreement will be signed by both sides at such time as all details of the proposed exhibition have been agreed upon through sufficient discussions.

At the request of the U.S. side, the CCPIT assisted the U.S. side in arranging meetings with concerned departments and enterprises relating to the proposed exhibition.

For the U.S. side:

For the Chinese side:

Dated: February 27, 1981

The Chinese and English versions of the signed Memorandum of Understanding.

To accommodate the American Exhibition in Beijing, two main air-supported structures, designed by I. M. Pei, would be erected on the site. The welcoming pavilion, a 670-foot long pneumatic tube, 55 feet in diameter, would carry visitors on moving sidewalks past exhibits and films. The second pneumatic structure, with a translucent triangular roof formed of woven synthetic fiber, would rise approximately ten stories at its highest point. It would enclose an exhibition hall that would contain a theatre complex, a tele-communications center with transponder facilities, and major exhibits about science, technology and culture.

The planners estimated that the cost of the entire US exhibition project in Beijing, including its design, fabrication and installation, as well as its operating costs for four months, would amount to $ 25,500,000. Funding was to be shared jointly by the US Government and the private sector. A number of American corporations, including ITT and Xerox, promised to support the American effort in Beijing. However, government support for the project was tepid at best. The political will was not there, nor was the money.

Ground Level Plan

0 32 m

Key to Exhibits

1 Applied Science and Research
2 Information and Electronics
3 Breakthrough Materials
4 Space Odyssey: The Space Shuttle
5 Agriculture
6 Sports and Recreation
7 Industrial Design
8 Visual Communications
9 Innovations in Medical Technology

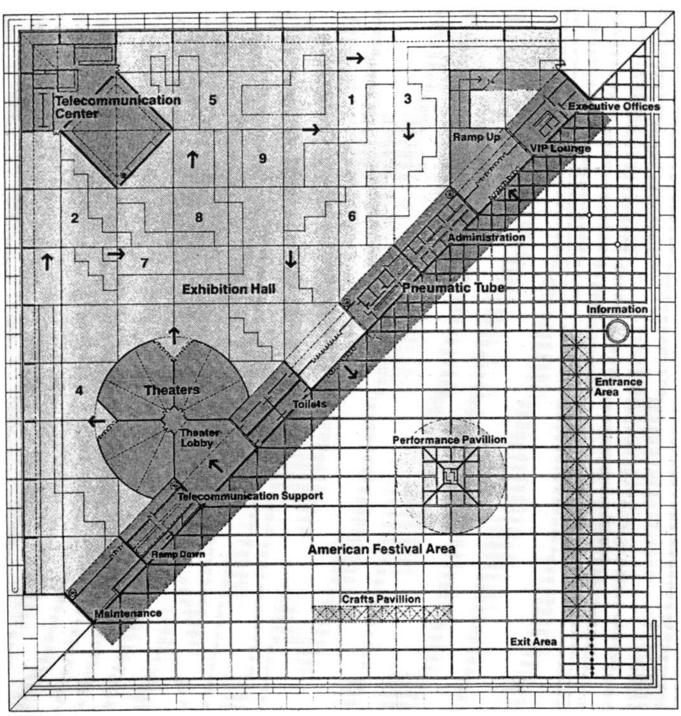

Plan views of the two-level US exhibition site.

"I want to say this about exhibitions: difficult as relationships had been between the United States and the Soviet Union, we made contact with other human beings. If the globe is ever to stop spending most of its energy fighting each other, and calling each other evil, the only way is to have some contact. It's got to go on. It's our only hope."

George Feifer, American guide

Upper Level Plan

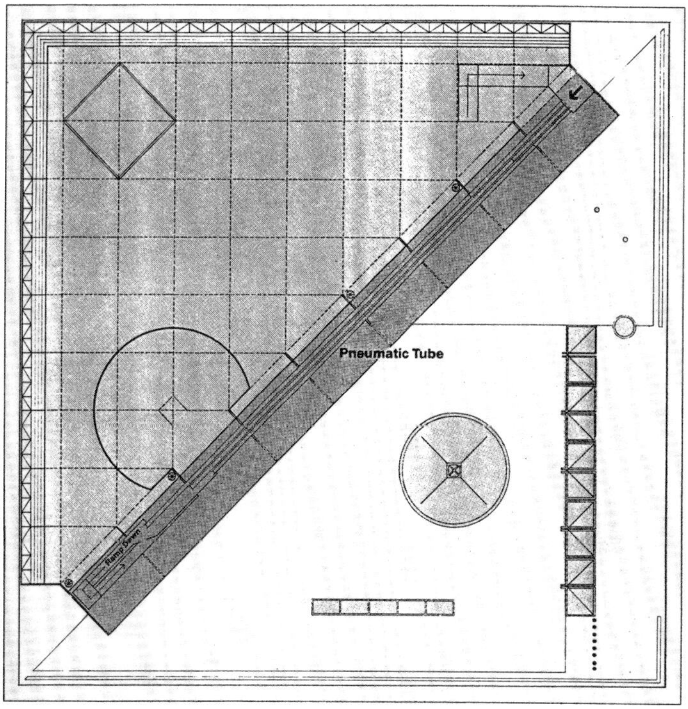

> Architect's model of I. M. Pei's design for the American exhibition in Beijing.

THE ABOLITION OF THE USIA

The USIA was abolished in 1999 after years of diminishing congressional funding, in part as the result of a four year battle between the Clinton Administration and Republican Senator Jesse Helms. Helms, the powerful Chairman of the Senate Foreign Relations Committee, was a long-time foe of USIA. Many of USIA's functions were transferred to the State Department. It could be argued that the USIA, a creation of the tensions of the Cold War, was no longer needed in the post-Soviet world. But this would be a mistaken argument. The need for America to communicate with the world remained just as important as ever, and the need for feedback into Washington of foreign opinion about the US and its policies was no less critical.

According to US diplomat Neil R. Klopfenstein, "The integration of USIA into the State Department ... insured that American public diplomacy at the beginning of the 21st century would be defined by more policy advocacy and less promotion of mutual understanding." While some of the functions of USIA were transferred to the State Department, its collective experience was not, nor, more importantly, was its independent voice. This break in continuity is recognised by many as serious. Eugene "Rocky" Staples, who was the Deputy General Manager of the American Exhibition in Moscow, writes in "Old Gods, New Nations": "As a new century uneasily begins, (America is) enormously, frighteningly powerful, yet at the same time perilously friendless and vulnerable." Although the end of the Cold War eventually resulted in the abolition of the United States Information Agency on October 1, 1999, the impetus within the US government to "capture hearts and minds" had atrophied over a period of several years.

In 2007, members of Congress proposed recreating the Agency: America today needs more than ever to engage in "the promotion of mutual understanding," especially in the Islamic world. But the value of an agency such as USIA is in its collective memory, developed through its historic experience and its accumulation of knowledge: it is not something one can jump-start.

One of the technical pretexts for abolishing USIA was that it had, in effect, violated the Smith-Mundt Act's prohibition against operating within the United States by opening websites, which could of course be accessed by US residents.

The gap between the rapid advance of new technologies and official policies has probably never been greater than it is today, and this gap is at the center of a vigorous debate within the US about how best to conduct public diplomacy in light of today's security concerns and complex global challenges.

"The really crucial link in the international communications chain is the last three feet, which is best bridged by personal contact—one person talking to another."

Edward R. Murrow, Director, USIA, 1961–1964

"A Russian uttered these parting words: You're not really devils, after all."

William Watts, American Embassy, Moscow, 1962

Currently, think tanks at George Washington University and at the University of Southern California are investigating how the use of new media might broaden opportunities for dialogue between American citizens and foreign publics, and the State Department has begun to explore the use of web portals (such as its "Ask the Ambassador" website) as part of a web-based public diplomacy outreach program. Exploration of ways in which new media might restore contact with audiences once served by USIA's libraries and cultural centers has yet to begin in earnest, however.

A Council on Foreign Relations taskforce has also made an interesting suggestion about how to address the public diplomacy needs of the 21st century. It has recommended the creation of two new organizations: one, headed by a presidential appointee, would coordinate public diplomacy programs and policies for the White House, and a second, non-profit Corporation for Public Diplomacy would help mobilize the private sector. Staples, today a member of the Council, writes that, "The final form of a revamped public diplomacy effort is less important than the brainpower, leadership, and money devoted to it. Its success will ultimately depend on the commitment and the policy leadership it gets from the president."

A NEW PUBLIC DIPLOMACY INITIATIVE

In reading reports and documents of the Cold War period, it is clear that the stalemate around a horrific nuclear technology served to maintain the two opposed superpowers in their respective spheres of influence: these documents also provide insight into a sense of urgency and engagement for cold warriors that recall the memoirs of the Second World War. For those who lived through it, it was a real experience of combat, and a combat in which all weapons, except the nuclear ultimate, could be used. For the USIA the chosen weapon was information outreach, and part of its arsenal of communication was the medium of exhibitions, designed to illuminate, inform and influence as wide an audience as possible. Given the often limited funding and resources available to USIA generally, and its exhibition programs in particular, it was a hard and demanding role, but not without excitement, humor and real pleasure in achievement.

The role of the guides at these international exhibitions was of particular importance, especially those that were held in the Soviet Union and Soviet bloc countries. The opportunity for Soviet citizens to meet and talk with a real American, in their own language, was an extraordinary one at the time, and one that was very effective in breaking down the stereotypes and preconceptions fostered on both sides by fear and misunderstanding. Today, film producer Linda Gottlieb, who was a guide at the American National Exhibition in Moscow in 1959, sums up the sense of many who participated in these giant cultural exhibitions, asking a question central to the concerns of today's public diplomacy and design communities: "Those of us lucky enough to have been in Moscow fifty years ago know the vivid, life-changing power of person-to-person diplomacy. We also understand why it is impractical in today's world. But if we cannot send guides to talk to millions of people around the world any more, how can new media create imaginative ways of reaching out to those millions of people to foster better mutual understanding among us all?"

Today, social networking websites offer new opportunities to connect with people in other parts of the world, but with the difference that they are virtual, not real: you do not know to whom you are talking or what their real status is. The world's fairs may no longer be the ideal route to illumination and information; but there is a need for all governments to have a representational presence in the global community through which to foster and maintain dialogue with old friends and new enemies alike.

New media, such as the worldwide web and digital television, offer an overwhelming range of possibilities to communicate with others, but need a strong voice and a determined presence to be heard and seen there. In seeking how to find a voice in today's still inchoate Web, Thomas Geismar observes, "The question is, how do you get beyond all the noise?" The British Broadcasting Corporation's news website is one that has. But the prestige of the BBC website is the fruit of the long success of its broadcasting reputation, especially the radio-based World Service. Again, long-term investment has paid off in a new medium. As with a world's fair pavilion, the return on such an investment cannot be calculated. But the effort is worth it.

Ivan Chermayeff proposes a re-evaluation of what an exhibition might be in light of today's changed realities: "To be interesting [today's exhibitions] would need to be relatively sophisticated. With today's budgets so vastly larger than they were during the Cold War years, perhaps the government today should focus on designing projects that are the opposite of what was done for Cold War Expos—small, unpretentious and permanent."

All agree that any new US government initiative directed at expanding mutual understanding must have the widest possible access to the most proficient new media practitioners. And, it goes without saying, such a new initiative must take its forms and functions not just from its antecedents, but from the circumstances and challenges of today. The potential for such new information outreach initiatives is as exciting and challenging today as it was at the start of the Cold War. And again, perhaps no one can tell what the consequences would be.

"The 1959 American Exhibition in Moscow became the metaphor for so many of us on the nature of public diplomacy in the Cold War. Today, we need innovative thinkers to find new equivalents to the Moscow Exhibition that can be delivered by means of new media."

"The nature of the new problems of communications and story telling that the US faces today, with new adversaries and old friends alike, provides an exciting opportunity to look forward to the public diplomacy of the future."

Ambassador William H. Luers, President, United Nations Association of the United States of America

INDEX OF PROPER NAMES

BIBLIOGRAPHY

Abercrombie, Stanley. *George Nelson: The design of modern design.* Cambridge, Mass., London: MIT Press, 1995.

Allen, James Sloan. *The romance of commerce and culture: capitalism, modernism, and the Chicago-Aspen crusade for cultural reform.* Chicago: University of Chicago Press, 1983.

Andrew, Christopher M. and Vasili Mitrokhin, *The sword and the shield: the Mitrokhin Archive and the secret history of the KGB.* New York: Basic Books, 1999.

Arndt, Richard T. *The first resort of kings: American cultural diplomacy in the twentieth century.* Dulles, Va.: Potomac Books, 2005.

Banham, Reyner, ed. *The Aspen Papers; twenty years of design theory from the International Design Conference in Aspen.* New York: Praeger, 1974.

Behrman, Greg. *The most noble adventure: the Marshall Plan and how America helped rebuild Europe.* New York: Free Press, 2008.

Belmonte, Laura A. *Selling the American way: national identity, propaganda, and the Cold War, 1945/1959.* Philadelphia: University of Pennsylvania Press, 2008

Black, Misha. *Exhibition Design.* London: Architectural Press, 1950.

Blake, Peter. *No place like Utopia.* New York: Knopf, 1993.

Buell, Frederick. *From apocalypse to way of life: environmental crisis in the American century.* New York: Routledge, 2003.

—. *National culture and the new global system.* Baltimore: Johns Hopkins University Press, 1994.

Caute, David. *The dancer defects: the struggle for cultural supremacy during the Cold War.* Oxford; New York: Oxford University Press, 2003.

Chanzit, Gwen Finkel. *From Bauhaus to Aspen: Herbert Bayer and modernist design in America.* Boulder, Col.: Johnson Books, 2005.

Cohen, Lizabeth. *A consumers' republic: the politics of mass consumption in postwar America.* New York: Knopf, 2003.

Colomina, Beatriz, Anne Marie Brennan, and Jeannie Kim, eds. *Cold war hot houses: inventing postwar culture, from cockpit to playboy.* New York: Princeton Architectural Press, 2005.

Colomina, Beatriz. *Domesticity at war.* Cambridge, Mass.: MIT Press, 2007.

Chomsky, Noam [et al.] *The Cold War & the university: toward an intellectual history of the postwar years.* New York: New Press, 1997.

Cook, Blanche Wiesen. *The declassified Eisenhower: a divided legacy.* New York: Doubleday, 1981.

Cull, Nicholas John. *The Cold War and the United States Information Agency: American propaganda and public diplomacy, 1945–1989.* Cambridge; New York: Cambridge University Press, 2008.

Dernie, David. *Exhibition design.* New York: W. W. Norton, 2006.

Dizard, Wilson P. [Jr]. and S. Blake Swensrud. *Gorbachev's information revolution: controlling glasnost in a new electronic era.* Boulder, Colo: Westview Press, 1987.

Dizard, Wilson P., Jr. *Inventing public diplomacy: the story of the U.S. Information Agency.* Boulder, Colo.: Lynne Rienner Publishers, 2004.

—. *The coming information age: an overview of technology, economics, and politics.* New York: Longman, 1982.

—. *Old media, new media: mass communications in the information age.* New York: Longman, 1994.

Dunlap, Henry A. and Hans N. Tuch. *Atoms at your service.* New York: Harper, 1957.

Elder, Robert E. *The information machine; the United States Information Agency and American foreign policy.* Syracuse, N.Y.: Syracuse University Press, 1968.

Fox, Richard Wightman and T. J. Jackson Lears, eds. *The Culture of consumption: critical essays In American history, 1880–1980.* New York: Pantheon Books, 1983.

Fuller, R. Buckminster. *Critical path.* New York: St Martin's Press, 1981.

—. *Guinea pig B: the 56 year experiment.* Clayton, CA. Critical Path Publishing, 1983.

—. *Ideas and integrities, a spontaneous autobiographical disclosure.* Englewood Cliffs, N. J.: Prentice Hall, 1963.
Fursenko, Aleksandr and Timothy Naftali. *Khrushchev's cold war: the inside story of an American adversary.* New York: Norton, 2006.

Gaddis, John Lewis. *The Cold War: a new history.* New York: Penguin Press, 2005.

Gorman, Michael John, *Buckminster Fuller: designing for mobility.* Milan: Skira, 2005.

Haddow, Robert H. *Pavilions of plenty: exhibiting American culture abroad in the 1950s.* Washington: Smithsonian Institution Press, 1997.

Hixson, Walter L. *Parting the curtain: propaganda, culture and the Cold War, 1945–1961.* Houndmills: Macmillan, 1997.

Hobsbawm, Eric. *The age of extremes: a history of the world, 1914–1991.* London: Michael Joseph, 1994

Huntington, Samuel P. *The clash of civilizations and the remaking of world order.* New York: Simon & Schuster, 1996.

Jones, Caroline A. *Eyesight alone: Clement Greenberg's modernism and the bureaucratization of the senses.* Chicago: University of Chicago Press, 2005.

Judt, Tony. *Postwar: a history of Europe since 1945.* New York: Penguin Press, 2005.

Krausse, Joachim, Claude Lichtenstein, eds. *Your private sky: R. Buckminster Fuller, the art of design science.* Translation by Stephen Lindberg, Julia Thorson. Baden: Lars Müller Publishers, 1999.

Krenn, Michael L. *Fall-out shelters for the human spirit: American art and the Cold War.* Chapel Hill: University of North Carolina Press, 2005.

Kuisel, Richard F. *Seducing the French: the dilemma of Americanization.* Berkeley: University of California Press, 1993.

Lasswell, Harold D. *Propaganda technique in World War I.* Cambridge, Mass.: MIT Press, 1971.

Margolin, Victor and Richard Buchanan, eds. *The idea of design.* Cambridge, Mass.: MIT Press, 1995.

Marks, Robert W. *The Dymaxion world of Buckminster Fuller.* New York: Reinhold Publishing Corp., 1960.

Marling, Karal Ann. *As seen on TV: the visual culture of everyday life in the 1950s.* Cambridge, Mass.: Harvard University Press, 1994.

Martin, Manuel and Anna Noguera, eds. *El Cadaqués de Peter Harnden i Lanfranco Bombelli.* Col.legi d'Arquitectes de Catalunya, 2002.

Mattie, Erik. *World's fairs.* New York: Princeton Architectural Press, 1998.

Meikle, Jeffrey L. *Design in the USA.* Oxford; New York: Oxford University Press, 2005

—. *Twentieth century limited: industrial design in America.* Philadelphia: Temple University Press, 1979.

Morgan, Conway Lloyd. *20th Century design: a reader's guide.* Oxford; Boston: Architectural Press, 2000.

Nelson, George. *Building a new Europe: portraits of modern architects: essays by George Nelson, 1935–1936.* Introduction by Kurt W. Foster; Forward by Robert A. M. Stern. New Haven: Yale University Press in association with Yale University School of Architecture, 2007.

Nelson, George, ed., with an introduction. *Display.* New York: Whitney Publications, 1953.

Nelson, George. *Tomorrow's house; a complete guide for the home-builder.* New York: Simon and Schuster, 1945.

Neuhart, John, Marilyn Neuhart, Ray Eames, eds. *Eames design: the work of the office of Charles and Ray Eames.* New York: H. N. Abrams, 1989.

Osgood, Kenneth. *Total Cold War: Eisenhower's secret propaganda battle at home and abroad.* Lawrence: University of Kansas, 2006.

Papernyĭ, Vladimir. *Architecture in the age of Stalin.* Translated by John Hill and Roann Barris in collaboration with the author. New York: Cambridge University Press, 2002.

Pulos, Arthur. *The American design adventure, 1940–1975.* Cambridge, Mass.: MIT Press, 1988.

Pawley, Martin. *Buckminster Fuller.* London: Trefoil Publications, 1990.

Richmond, Yale. *Cultural exchange & the Cold War.* University Park: Pennsylvania State University Press, 2003.

——. *Practicing public diplomacy: a Cold War odyssey.* New York: Berghahn Books, 2008.

Rydell, Robert W. *All the world's a fair: vision of empire at American International Expositions, 1876–1916.* Chicago: University of Chicago Press, 1984.

History of Photography, vol. 29, no. 4. Winter 2005. "The Family of Man" issue.

Sandeen, Eric J. *Picturing an exhibition: the family of man and 1950s America.* Albuquerque: University of New Mexico Press, 1995.

Saunders, Frances Stonor. *The cultural cold war: the CIA and the world of arts and letters.* New York: New Press, 2002.

——. *Who paid the piper?: the CIA and the cultural Cold War.* London: Granta Books, 1999.

Schiller, Herbert I. *Communication and cultural domination.* White Plains, N.Y.: International Arts and Sciences Press, 1976.

——. *Culture Inc: the corporate takeover of public expression.* New York: Oxford University Press,1989.

——. *Mass communications and American empire.* Boulder, Colo.: Westview Press, 1992.

——. *The mind managers.* Boston: Beacon Press, 1973.

Schwartz, Richard A. *Cold War culture: media and the arts, 1945–1990.* New York: Checkmark Books, 2000.

Sorensen, Thomas C. *The word war; the story of American propaganda.* Introduction by Robert F. Kennedy. New York: Harper & Row, 1968.

Staples, Eugene. *Old Gods, New Nations: a memoir of war, peace and nation building.* iUniverse, 2006.

Walsh, Victoria. *Nigel Henderson: parallel of life and art.* Foreword and afterword by Peter Smithson. London: Thames and Hudson, 2001.

Weart, Spencer R. *Nuclear fear: a history of images.* Cambridge, Mass.: Harvard University Press, 1988.

Weiner, Tim. *Legacy of ashes: the history of the CIA.* New York, Doubleday, 2007.

Whitfield, Stephen J. *The culture of the cold war.* Baltimore: Johns Hopkins University Press, 1991.

Wilford, Hugh. *The mighty Wurlitzer: how the CIA played America.* Cambridge, Mass.: Harvard University Press, 2008.

PHOTO CREDITS

All photos are from the United States Information Agency unless otherwise noted:

KEYSTONE/AP Photo: Associated Press: 206, 207.
Courtesy, Marilyn Harris Bernard: top: 220, top left: 222, top: 245.
Peter Blake: top: 194, 228, 229.
Courtesy, Chermayeff & Geismar, LLC: 124, 125, 285, 294, 295, 296, 297, 298, 299, 300, 309, 313, 314, 315, 316, 319, 320, 321, 323, right: 325, 326, 327, 328, 329, 330, 331, 332, 333, 334, 335, 336, 337, 338, 339, 351, 356, 357, top: 361, 370, 371, 372, 373, 375, 376, 377, 378, 379, 380, 381, 382, 383, 384, 385, 387, 388, 389, 390, 391, 392, 393.
Courtesy, Marie Delikat Ciliberte: top left, bottom: 302.
Lucia DeRespinis: 185, 186, 187, 190, 191, top: 200, top: 201, 218, 219, bottom: 246.
© 2008 Eames Office LLC (www.eamesoffice.com) page 180, 181.
US National Archives and Records Administration (NARA): 64, 115, 122, 123, 128, 134, 137, bottom: 139, 140, 141, 142, 143, 144, 145, 146, 147, 148, 149, 150, 151, 164, 168, 169, 172, 173, 178, top: 183, 184, bottom: 188, 205, 210, bottom: 216, top left: 217, lower right: 240, 241, bottom: 245, 317, 344.
Courtesy, Philip George: 9, 12, 13, 16, 17, 19, 21, 25, 28, 29, 30, 31.
Courtesy, Linda Gottlieb: top right: 224.
Masey Archives: 66, 67, 160, 161, 247, 347, 374, 404, 405.
Mary Lou Leach Masey: 225.
Leo Lionni: 130, 131, 132, 133, 135, 136, top: 139.
Courtesy, George Malko: top left and bottom: 224, top: 238, 401.
The National Building Museum, Washington D.C., Lautman Collection: 253–282.
Courtesy, Jacqueline Nelson: top: 216, 244.
Courtesy, I. M. Pei: 406, 407, 408, 409.
Courtesy, Donald L. Richter: 171.
Courtesy, Tania Sochurek: lower left, 240.
U.S. Library of Congress, AP World Wide Press: 348, 349.
Courtesy, Vitra, Nelson Archives: Page 195.
Sheldon Weinig: top right, bottom: 217, bottom: 220.

ACKNOWLEDGEMENTS

This book would not have been possible without the enthusiasm, which was unstinting in the development of the project, the energy, which was endless for the research phase, and encouragement, which was profound during the assembly of the final work, of Beverly Payeff-Masey: the authors are forever in her debt.

A very special acknowledgment must go to those who are no longer living whose talents and contributions made the stories and projects in this book possible: Peter Blake, Lanfranco Bombelli, Abraham Brumberg, Samuel Brody, Jefferson Davis Brooks III, Terrence Catherman, Howard L. Chernoff, Lewis Davis, Rudolph de Harak, John Dixon, Charles Eames, Ray Eames, Milton "Micky" Fredman, R. Buckminster Fuller, David Geiger, William J. Handley, Peter G. Harnden, Emily Jones, Foy D. Kohler, Leo Lionni, Wallace "Pic" Littell, Harold "Chad" McClellan, Jane Fisk Mitirachi, George Nelson, Mary Painter, Charles Shaw, Robert Sivard, John. J. Slocum, Edward Durell Stone, Ted M. G. Tanen, Llewelyn E. Thompson, Charles Vetter and Abbott Washburn.

The authors are indebted to many who have generously contributed their time, insights, talents and other assistance to our effort to tell the stories of the great exhibitions of the Cold War era: Gourgen Assaturian, Helen Atlas, David Barber, Arthur Thomas Beach IV, Marilyn Harris Bernard, Casey Blake, Irma Weinig Bonilla, Chrysanthe B. Broikos, the Temple Hoyne Buell Center for American Architecture at Columbia University, Robert S. Byrnes, Alan Carter, Ivan Chermayeff, Peter Chermayeff, Marie Delikat Ciliberti, George R. Clack, Charles H. Clark, Lucia DeRespinis, Wilson P. Dizard, Donna Driscoll, the Eames Office, George Feifer, Peter Floyd, Genevieve Fong, Nancy Friday, Thomas Geismar, Philip George, Linda Gottlieb, Janice Grenci, Jane Helms, Sylvia Juran, Nina L. Khrushcheva, Jennifer Minjung Kim, Ray Komai, Tami Komai, Joseph I. Krene, Tab Lewis, Louis Mannie Lionni, Robert C. Lautman, the archivists at the Library of Congress, William H. Luers, Elizabeth Malko, George Malko, Martin Manning, Phyllis Montgomery, the archivists at the National Archives and Records Administration, the National Building Museum, Nick Natanson, Jacqueline Griffith Nelson, John Neuhart, Marilyn Neuhart, Christina Blake Oliver, Vladimir Paperny, Mary Patzer, Jane Pavitt, I. M. Pei, Terry Rankine, Donald L. Richter, Brendan Rooney, Shoji Sadao, Alan Schwartzman, Kate Scott, Jaime Snyder, Tania Sochurek, Zoë Sheehan Saldana, Eugene "Rocky" Staples, George Stevens, Jr., David Sutton, Tomas Tolvaisas, Neil Thompson, Hans N. "Tom" Tuch, Dana Twersky, Yasuo Uesaka, William Watts, Donald M. Wilson and Sheldon Weinig.

Finally, a very special note of thanks from both authors must go as well to our interns Angela Fabregas, Joann Huang and Franklyn Roa of Bernard Baruch College, and to Michael Corris, Geraint Cunnick, David Smith and colleagues at the Newport School of Art, Media and Design. The effort and enthusiasm of Louise Stein, Michael Furrer and others at Lars Müller Publishers is also much appreciated.

Jack Masey, Conway Lloyd Morgan

Cold War Confrontations
US Exhibitions and Their Role
in the Cultural Cold War

Design/Typography: Integral Lars Müller/Lars Müller and Michael Furrer
Design Assistance: Lea Pfister
Copyediting: Louise Stein
Production: Marion Plassmann
Printing and Binding: EBS Bortolazzi STEI, Verona/Italy

Printed in Italy

© 2008 Lars Müller Publishers
No part of this book may be used or reproduced
in any matter whatsoever without any written
permission except in the case of brief quotations
embodied in critical articles and reviews.

Lars Müller Publishers
Baden, Switzerland
www.lars-muller-publishers.com

ISBN 978-3-03778-123-4